More praise for

WORLD IN THE BALANCE

Zòcalo Public Squa

"Precision and fastidiousness— stem of measurement might seem a ophher Robert P. Crease makes clear in *World in the Balance*, it was anything but. From prehistory to the present, Crease ties humanity's search for precision to the history of nations and of ideas. . . . [A] fascinating book." —Arthur I. Miller, *New Scientist*

"Mr. Crease . . . shows that the story of metrology, not obvious material for a page-turner, can in the right hands make for a riveting read."

—*Economist*

"Crease, philosophy professor at Stony Brook University, proves an adept explainer of both physical science and metaphysical dilemmas—such as, do our choices in how and what we measure both reveal and shape what we find important?" —Kate Tuttle, *Boston Globe*

"Educational and often entertaining. . . . A sweeping survey of the history of measurement and the search for universal and absolute standards, from ancient China up to practically yesterday."

—*Wall Street Journal*

"Anyone who doubts the romance and history to be found in a meter stick will find this book a revelation. . . . Through entertaining anecdotes and history, Stony Brook philosophy chair and *Physics World* columnist Crease (*The Great Equations*) ably reveals our modern world as a 'metroscape' shaped by the things we measure and the way we measure them." —*Publishers Weekly*

"Intellectual adventure doled out by the pound, kilogram, and metric ton."
—Bryce Christensen, *Booklist*

"An intricate, surprising history of 'metrology,' the science of measuring things."
—Josh Rothman, *Braniac*, Boston.com

"An engaging story. [Crease] ably succeeds by communicating information through well-told narratives and adding appetizing tidbits from real-life situations, such as changing official bell tones in China or finding the true meter measure in a vault. . . . Superbly presented and will prove accessible to readers with any and all levels of background knowledge. . . . Highly recommended for all readers interested in history and anyone interested in how we see and interact with our world."
—Eric D. Albright, *Library Journal*

"A fascinating story, and all the more so because it is full of optimism. . . . Crease in his role as philosopher knows that measuring is more than just applying precision, but is a human endeavor that must always be tied to human enthusiasms and activities."
—Rob Hardy, *Commercial Dispatch*

"Robert P. Crease's *World in the Balance: The Historic Quest for an Absolute System of Measurement* globe-trots exuberantly from Han Dynasty China to Victorian England to mid-20th-century Africa to explain why we measure things the way we do, answering fundamental questions about time, space, and science in the process."
—*Zócalo Public Square*

"There's more to the meter than a metal stick. Crease, a physicist and philosopher, traces the rise of the metric system, telling a colorful tale of global conquest driven by kings, revolutionaries, polyglots and privateers—and ultimately scientists looking for rulers that could outlast any physical object."
—*Science News*

"Crease provides a solid explanation of how something so arbitrary can be made truly 'universal.' Scientific history that looks beyond the facts and figures to their influence on everyday life." —*Kirkus Reviews*

"With his typical big-picture vigor, Robert P. Crease guides us on one of history's greatest philosophical adventures—one that encompasses the French Revolution, surrealism, and the speed of light. *World in the Balance* takes the seemingly mundane questions we unthinkingly ask dozens of times a day and reveals them to be thrillingly profound."
—Richard Panek, author of *The 4% Universe: Dark Matter, Dark Energy, and the Race to Discover the Rest of Reality*

"This curious—and curiously interesting—work delves into history, philosophy, science, and art as it explores why we measure what we do, and what that measurement means. As with the work of James Gleick, it's both rigorous and deeply enjoyable to read." —Andrea Barrett, author of *Ship Fever: Stories* and *The Air We Breathe*

"Metrology—the scientific study of measurement, the subject of *World in the Balance*—is a topic that seems small at first look but is in fact a globe-spanning epic with a cast of characters ranging from John Quincy Adams and the brilliant Swiss-American lunatic Ferdinand Hassler to bra models in Manhattan, painters in the Chinese Cultural Revolution, and the pyramid-obsessed Scottish astronomer Piazzi Smyth. This story is so little known because telling it well requires the combined talents of a scientist, a historian, and a philosopher. Fortunately for readers, Robert Crease can wear all of these hats. *World in the Balance* has something of the flavor of Simon Winchester's *Atlantic* and Dava Sobel's *Longitude* but is entirely its own terrific book." —Charles C. Mann, author of *1493: Uncovering the New World Columbus Created*

ALSO BY ROBERT P. CREASE

The Great Equations: Breakthroughs in Science from Pythagoras to Heisenberg

The Prism and the Pendulum: The Ten Most Beautiful Experiments in Science

Making Physics: A Biography of Brookhaven National Laboratory

The Play of Nature: Experimentation as Performance

The Second Creation: Makers of the Revolution in Twentieth-Century Physics, by Robert P. Crease and Charles C. Mann

Peace & War: Reminiscences of a Life on the Frontiers of Science, by Robert Serber with Robert P. Crease

J. Robert Oppenheimer: A Life, by Abraham Pais with Robert P. Crease

TRANSLATED BY ROBERT P. CREASE

What Things Do: Philosophical Reflections on Technology, Agency, and Design, by Peter-Paul Verbeek

American Philosophy of Technology: The Empirical Turn, edited by Hans Achterhuis

WORLD IN THE BALANCE

The Historic Quest for an Absolute System of Measurement

ROBERT P. CREASE

W. W. NORTON & COMPANY
NEW YORK | LONDON

Printed in the United States of America
First printed as a Norton paperback 2012

For information about permission to reproduce selections from this book, write to Permissions, W. W. Norton & Company, Inc., 500 Fifth Avenue, New York, NY 10110

For information about special discounts for bulk purchases, please contact W. W. Norton Special Sales at specialsales@wwnorton.com or 800-233-4830

Manufacturing by Courier Westford
Book design by Marysarah Quinn
Production manager: Julia Druskin

Library of Congress Cataloging-in-Publication Data

Crease, Robert P.
World in the balance : the historic quest for an absolute system of measurement / Robert P. Crease. — 1st ed.
p. cm.
Includes bibliographical references and index.
ISBN 978-0-393-07298-3 (hardcover)
1. Weights and measures—History. 2. Measurement—History.
3. Metrology. I. Title.
QC83.C74 2011
530.8'1—dc23
2011029026

ISBN: 978-0-393-34354-0 pbk.

W. W. Norton & Company, Inc.
500 Fifth Avenue, New York, N.Y. 10110
www.wwnorton.com

W. W. Norton & Company Ltd.
Castle House, 75/76 Wells Street, London W1T 3QT

1 2 3 4 5 6 7 8 9 0

For Stephanie,
beyond measure

CONTENTS

WORLD IN THE BALANCE

INTRODUCTION
THE NOONDAY CANNON

For centuries, a remote seaside village on a distant foreign shore kept time by the cannon that a local military base fired at exactly noon from its perch atop a nearby hill. This was long before the days of the Internet—before even television and radio—and in the village the sound of the noonday cannon was a phenomenon as natural as the rising and setting of the sun, a regularity that celebrated the day and divided morning from afternoon. The noonday cannon shaped the stability and pace of the villagers' lives, and was used to plan everything from business meetings to conducting clandestine affairs.

According to this old legend, a teenage boy wondered how the cannon knew to go off at exactly noon. One day he climbed the hill and asked the artilleryman how he fired the cannon. The artilleryman smiled at the boy. He shot the cannon by order of the commanding officer, whose duties included procuring the most accurate watch obtainable and keeping it carefully synchronized. The youth then approached the commanding officer, who proudly showed him the finely made and precise timepiece. How was it set? On his weekly walk into town, the commander said, he always took the same route, which carried him past the shop of the town watchmaker. He would stop and synchronize his watch with the large and venerable clock in the watchmaker's window, one that many others in town also used to set the time.

The next day, the youth visited the watchmaker and asked how he set the large clock in the window. "By the only reliable way anyone around here has ever had," the watchmaker replied. "I set it by the noonday cannon!"

The story of the noonday cannon in the seaside village captures the habitual way we rely on measures. A measure, most simply, is a standard or mark against which we gauge or evaluate something; in the village, people used the cannon shot to mark the time of the morning from the time of the afternoon. Once a measure exists, it is irresistible to take it for granted and suppose it always existed. Measures become part of the contour of things, seeming to belong to the world itself. But each standard or mark first entered the world as the result of a human decision. Nature provides no measuring rods, no scales, and—though days and years recur regularly—no convenient zero points for setting time. We fashion these ourselves, with the aid of things like sundials and watches. Often we check our measures against other standard measures we've fashioned, or assume that there are such standard measures somewhere; as in the cosmological myth that the earth is supported by an elephant that stands on turtles, there's always a final turtle. The result can be arbitrary: if the remote seaside village's watches suddenly all broke and the townspeople had to invent a new midday marker, it would certainly be different—12:04, 11:47, 1:28—but the difference would not matter. The result can be both arbitrary and circular: we define a measure by some element of the world, and that element by the measure. We call noon the moment the cannon fires, and fire the cannon at noon.

The arbitrariness and the circularity simply reflect the typical way that people select measures: we improvise something accessible from our surroundings. Cultures all over the globe have been improvising measures since the dawn of history. During the birth of modern science in the 1600s and 1700s scientists in France tried to develop a universal system of measurement that could be shared by all countries and that would be tied to unchanging features of nature. They succeeded at the first goal but not the second. Finally, half a century ago, an international organization of scientists succeeded in tying one unit of measurement, length, to a natural phenomenon, light. Other measures, such as time, soon were also tied to natural phenomena. Today, the only fundamental or "base" measure not tied to a natural phenomenon is mass. The mass unit is defined by a lump of metal sitting in a vault in a laboratory outside Paris. But its days as the ultimate stone governing mass measurement—as the final turtle—are numbered. Today, a new generation of scientists is on the verge of taking yet another step, tying all basic units of measure together—including mass—and defining them in terms of physical constants, to produce an "absolute" system of measurement. For the first time in history, if all basic standards were somehow lost, they could be recovered and the world would have exactly the same measurement standards as before. This book is about how that happened.

CHAPTER ONE

VITRUVIAN MAN

Daniel Defoe's novel *Robinson Crusoe* (1719) contains one of the most famous acts of improvising a measure of all time. Crusoe had been shipwrecked on a deserted island for 15 years when, strolling on the beach, he was "thunder-struck" to see "the print of a man's naked foot" in the sand. After having lived for years without encountering trace of another living human, Crusoe was "terrify'd to the last degree." He retreated to his cave, tormented for 3 days and nights by "wild ideas." Was it Satan's own foot? Tracks of cannibals? Could it have been Crusoe's own footprint, and his fears but delusions? He could think of only one way to go on: "I should go down to the shore again, and see this print of a foot, and measure it by my own." Returning to the beach, Crusoe set his foot alongside the print of the other. The footprint was larger than his—much larger. Thanks to that measurement, he was sure that the island had been visited by at least one person other than himself. This discovery transformed Crusoe's notions about his own safety and prompted him to fortify his cave dwelling, which he henceforth thought of as his "castle."[1]

IMPROVISED MEASURES

In many respects, Crusoe's act illustrates the phases of all acts of measurement. He needs information that he can get only by comparing a property of a familiar thing (the length of his foot) with the same property of another unknown thing (the mysterious sand print), either to discover more about that other thing or, as when "measuring out" something like seeds or liquids or lengths of wood, to put it to use. The basic amount is called the *unit*; here, the unit was the "Crusoe foot." Acts of measurement may be routine or complicated, require only eyeballing or elaborate instruments, and be executed poorly or well. In each case, we hope to get a better grip on the world through our measures. The outcome can change our world; it certainly did Crusoe's.

The human body was the first and oldest measuring instrument. Feet are accessible; everybody has one. Nearly every civilization once had a "foot" unit, often subdivided into "fingers." In ancient Greece, for instance, the foot-measure or *pous* was subdivided into 16 fingers or *dactyloi*; in China the foot-measure was called the *chi* and subdivided into *cun*. Other units of length related to the body include the finger, fingernail, human hair (a few thousandths of an inch in diameter), palm, hand (still used to measure horses), forearm (also "ell" or "cubit"), span, step, and pace (double step). "Fistful," "handful," and "pinch" are still used as cooking units, and Ethiopians used "earhole" to measure out medicines.[2] Time units tied to human life have included heartbeats, lifetimes, and generations. According to an old story, the Russian general Alexander Suvorov defined an *arshin*, a unit of length, as a soldier's step, to assure that the members of his army would never be without one. A thousand or so *arshin* made up a *verst*, a bigger unit of length about a kilometer long.

In the 1860s and 1870s, Thomas Montgomerie, a British surveyor working in India, deployed one of the most extensive and rigorous applications of body measures in mapping Tibet and other areas of central Asia. Many of these nations refused entrance to Westerners, and those who snuck in faced execution. To get around this, Montgomerie recruited two Himalayan cousins, Nain and Mani Singh, and spent two years teaching them surveying techniques. He trained them to walk with a pace of exactly 33 inches, or about 2,000 paces per mile, regardless of terrain. Disguised as Hindu lamas, or *pundits*, a Hindu term for "holy men," the Singhs kept track of distance with counters camouflaged as Buddhist rosary wheels. The wheels were equipped with 100 beads instead of 108, the traditional number on a rosary, and the Singhs dropped one bead every 100 paces. Using such methods, Nain in particular managed to measure large sections of Tibet including Llasa. The resulting information helped Montgomerie compile a map of Tibet and central Asia, which among other purposes assisted the British in their brutal invasion of Tibet four decades later.

Since ancient times, from China to the Americas, human cultures also improvised length and weight measures from grain and seeds, including rice, corn, millet, barley, and carob (whence our word *carat*). The weight and length of grain and seeds vary according to climate and swell up in rainy seasons, but they are easy to obtain and fairly solid. Authorities often defined these natural standards more specifically and consistently by insisting that the grains and seeds be taken in dry season and be of average size.

Accessibility is only one of three important properties of a measure. A second is appropriateness—a measure has to be the appropriate scale for the intended purpose. Improvising a measure won't work unless it is convenient to use, and a typical measure, say, is of a few units and not thousands or a thousandth. According to a famous old story, the twelfth-century British king Henry I introduced the "yard," or *ulna*, into British measures, decreeing it to be the length of his arm. As the art historian

Peter Kidson pointed out, something must be missing from this story. The British already had a system of length measures, and introducing a new one out of the blue would have bewildered merchants. They would have had to relate it to the ones they were already using. "It is hard to believe that anyone actually inventing new measures would have gone out of his way to create such confusion," Kidson wrote. "Then there was the problem of getting the new invention into circulation; and finally, though not least, the difficulty of persuading people to use anything new and unfamiliar." In the world of George Orwell's *1984*, whose dictators could force citizens to change their language overnight, it would be easy to get people to forget old measures and use novel ones, with everyday life continuing to run smoothly without disruption or inconvenience. In real life, this could not happen. If Henry was indeed responsible for introducing the new unit, Kidson continues, it would be because British cloth manufacturers, used to using the old Roman fathom of 12 feet, desperately needed some new unit about half that size to characterize their products, which they could readily connect with their existing measures. "The King's part in the business," Kidson concluded, "was not to invent but to fill a void."[3]

Besides being accessible and appropriate, a measure must also be assured, or sturdy and reliable enough for the intended purpose. Again, early cultures provide us with some strikingly original examples. In Eastern Europe, a practice among Jews who had lost a loved one was to light a candle on the jahrzeit, or anniversary of death. The candle was supposed to burn for 24 hours and was safeguarded in a container known as a jahrzeit (or yahrzeit) glass. Such precious items were never thrown away—inexpensive glass is a modern technology—and families kept theirs, reusing them as drinking glasses. The practice continued in America. In the title story of Philip Roth's first book, *Goodbye Columbus*, when the protagonist recalls his grandmother drinking "hot tea from an old jahrzeit glass," it is an effective detail in conjuring a transplanted ancestor. Those glasses were all about the same size, for all were made to hold the same amount of wax. They were thick and solid so they wouldn't crack under the heat of the candle. They were the right size to dole out recipe ingredients, making them a natural cooking measure. Your grandmother would say that a recipe called for so-and-so many "glasses" of water, flour, or matzoh meal—and you knew immediately how much, and knew you had the equipment to measure it out. Such recipes were at first passed on orally, and then written down by a later generation, typically as what a daughter remembered from her mother. The measure was approximate—Old Country recipes tended to be loose with quantities—but it worked.

Today, thanks to relative affluence and manufacturing flexibility, we have more agreeable and precise things to use as drinking vessels and measuring cups than the heavy jahrzeit glasses, and it is rare to see them used as measures. Still, the morphing of jahrzeit glasses from commemorative funereal item into measurement unit illustrates how measures arise. Measurement is a classic example of a technology in the ancient Greek sense of something that we do to "complete nature." Nature, the Greeks pointed out, has created human beings with needs (food, clothing, shelter) that nature itself does not fulfill or has only provided us with the raw materials for; it is up to us to finish the job and find or create something to satisfy our needs. In the case of the jahrzeit glass, the Eastern Europeans improvised; they took something created for one purpose and turned it to another. In other cases, human beings have to make something to satisfy this natural need for measures, fashioning things like rulers, scales, watches, and other instruments.

Nothing is inherently unscientific about using improvised measures, as long as they are accessible, appropriate, and assured. Wallace Sabine (1868–1919), a Harvard University physicist, was asked by

Harvard's president to fix the acoustics in the University's Fogg Art Museum, whose rooms were unpleasantly reverberant, and to come up with a quantitative measure of acoustical quality.[4] How could an elusive subject like reverberation be measured? Sabine decided to experiment with seat cushions. He conducted experiments in various rooms at Harvard, using seat cushions from the university's acoustically superior Sanders Theatre. Working between midnight and about 5 a.m., when quiet reigned over the campus, he and assistants removed all the cushions from the hall and, using a stopwatch, organ pipe, and a keen-eared assistant, measured how long a sound produced in Sanders continued to be audible with different numbers and positionings of cushions.[5] Sabine was able to derive the formula, $xy = k$, where x is the quantity of Sanders cushions, y the room's reverberation time, and k a constant. Shortly thereafter, Sabine established a famous and far-reaching formula relating reverberation time to absorption, volume, and surface area: $t = k/(a + x)$, where t is the reverberation time; k is a constant that depends on the volume of the room; a is the absorbing power of the walls, floor, and ceiling; and x is the absorbing power contributed by the furniture and audience. Sanders Theatre cushions are a kind of acoustical jahrzeit glass, something made for one purpose and transformed into a measuring unit for something completely different. In this other purpose, the cushions helped shepherd in a breakthrough that transformed how auditoriums, from classrooms to concert halls, were built the world over.[6]

Improvising measures has its limitations. Measures often have to cover a range of scales, for which any single measure is inadequate. Carpenters building a house require units that span fractions of an inch and feet and yards. Cooks have to handle units from pinches and tablespoons to cups and gallons. Sometimes a large-scale measurement unit is made by piling up quantities of a small one: the word "mile," for instance, stems from the Latin *milia passuum*, for a thousand paces. Other times, units are related to each other by patterns.

The Laws of Manu, an ancient Sanskrit text said to be the work of a legendary Hindu lawgiver and that dates from about 500 BC, outlines a pattern between measures commonly used in trading copper, silver, and gold:

> The very small mote which is seen when the sun shines through a lattice, they declare to be the least of all quantities and to be called a trasarenu (a floating particle of dust). Know that eight trasarenu are equal in bulk to a likshâ (the egg of a louse); three of those to one grain of black mustard (râgasarshapa) and three of the latter to a white mustard-seed. Six grains of white mustard are one middle-sized barley-corn and three barley-corns one krishnala.[7]

Doing the math, the krishnala (retti or rati seed) equals 1296 specks of dust.

We find another delightful illustration of improvised measure units in Eric Cross's entertaining novel *The Tailor and Ansty*, published in Ireland in 1942. The book gave such vivid voice to the bawdy sides of rural Irish life that it was banned when first published, and angry neighbors persecuted the couple (named by the title) on whose lives the novel was based. The Tailor is particularly fond of relating to his wife Ansty—and to anyone else within earshot—the wisdom of the old Irish before, as he puts it, "the people got too bloodyful smart and educated, and let the Government or anyone else do their thinking for them." Some of this wisdom involved measurement.[8]

Land area, the Tailor announces, used to be reckoned in "collops." The collop, based on the "carrying power" of land, "told you the value of a farm, not the size of it. An acre might be an acre of rock, but you know where you are with a collop." One collop was the area needed to graze "one sow or two yearling heifers or six sheep or twelve goats or six geese and a gander," while three collops were needed to graze a horse. A neighbor boasts of owning 4000 acres, but really only has "enough real land to graze four cows," the Tailor says. He surely exaggerates. Few people in his area own that much land, and even so 1000 acres in the west of Ireland, despite its famous bogs and rocky hills, would be more than enough to serve an average cow. But the Tailor is right that measuring in collops cuts the boast down to size. "The devil be from me! but the people in the old days had sense."

The Tailor also tells us that the old Irish were equally sensible about reckoning time. The basic unit was the lifespan of a rail, a type of small bird. The Tailor then translates from the Irish a pattern of units relating to the rail lifespan, illustrating a system of interconnected units:

A hound outlives three rails.
A horse outlives three hounds.
A jock outlives three horses.
A deer outlives three jocks.
An eagle outlives three deer.
A yew-tree outlives three eagles.
An old ridge in the ground outlives three yew-trees.

We have no need for other time units, the Tailor says, for three times the age of an old ridge is the age of the universe. He is wildly off in his estimate, for the age of the universe is unlikely to be the lifetime of a rail multiplied by 3^8. If the average rail lives, say, 10 years, that's only 65,610 years since the big bang, which is considerably shorter than current astronomical estimates of 14 billion years. Still, it is easy to appreciate the Tailor's point: old Irish units were "reckoned on the things a man could see about him, so that, wherever he was, he had an almanac."

The almanac of the world: what a wonderful expression to describe the origin of such measuring units! The Tailor doesn't impose any artificial or invented units on his world; he gets them from the world itself and its patterns.

PATTERNS OF UNITS

Patterns of units can have a special significance of their own. When the elements of a pattern had certain kinds of relations, the ancient Greeks called them proportionate or "symmetrical," from the Greek for "the coming together of measure." The human body provided a beautiful illustration of such symmetry. "A finely-shaped human body," writes the Roman architect and historian Vitruvius in his work *De Architectura*, written in the first century BC, has proportionate measurements:

> For the human body is so designed by nature that the face, from the chin to the top of the forehead and the lowest roots of the hair, is a tenth part of the whole height; the open hand from the wrist to the tip of the middle finger is just the same; the head from the chin to the crown is an eighth, and with the neck and shoulder from the top of the breast to the lowest roots of the hair is a sixth; from the middle of the breast to the summit of the crown is a fourth. . . . The length of the foot is one sixth of the height of the body; of the forearm, one fourth; and the breadth of the breast is also one fourth. The other members, too, have their own symmetrical proportions, and it was by employing them that the famous painters and sculptors of antiquity attained to great and endless renown.[9]

The measures of the human body, Vitruvius held, therefore have an aesthetic and religious significance, for their proportions mirror the cosmological order, reflect spiritual dimensions, embody harmony and perfection, and connect humanity with trans-human nature. For this reason, Vitruvius continued, "it was from the members of the body that they [the ancients] derived the fundamental ideas of the measures which are obviously necessary in all works." The most important of these measures were the *orguia* or fathom, the distance from fingertip to fingertip of the middle fingers of outstretched arms; the cubit or ell (fingertip to elbow), the foot, the palm, and the finger. When one such unit was used as the principal measure, Vitruvius called it a module. Many ancient buildings appear to have been laid out in grids using such a module. The platform of the Parthenon, for instance, was said to be 100 feet wide and 225 feet long, yielding an approximate measure of the Greek foot.

The Greeks sometimes inscribed these patterns into metrological reliefs. Here are two well-known examples:

The one above on the opposite page, made in Greece or western Turkey in the fifth century BC and now at the Ashmolean Museum in Oxford, England, shows the relationships among the orguia or fathom, cubit, foot, and fingers. Four cubits are in an orguia—though this is evidently the "royal" cubit, the one used inside the court, not the one used by ordinary citizens in the marketplace. The relief below it, made at Salamis in Greece in the fourth century BC and now at the Piraeus Museum, shows relationships among the orguia, cubit, foot, and span (tip of thumb to tip of little finger when the fingers are outstretched).[10]

Another example is Leonardo da Vinci's famous and much copied and caricatured drawing frequently referred to as the Vitruvian Man, for he evidently had the architect's passage in mind. Leonardo's Vitruvian Man displays how the proportions of the human body, and the units drawn from it, participate in an ideal of beauty. The Vitruvian Man shows us that the organization of measures can have symbolic and spiritual significance.

Already in prehistory, however, humans discovered the need, for many

METROLOGICAL RELIEF FROM GREECE OR WESTERN TURKEY IN THE FIFTH CENTURY BC, SHOWING THE RELATIONSHIPS AMONG VARIOUS PARTS OF THE MALE BODY.

METROLOGICAL RELIEF FROM SALAMIS IN GREECE IN THE FOURTH CENTURY BC.

purposes, to select a particular object to define a unit of measure—*a* foot, not yours or mine; *a* carat, not this one or that. This particular thing is called a *standard*. A standard is a sample of a particular quantity which we have chosen to specify as having the value 1 of that quantity. When a standard is created, it *embodies* the unit, giving it a specific, concrete identity as an artifact.[11]

Shifting from a measure taken from the almanac of the world to an

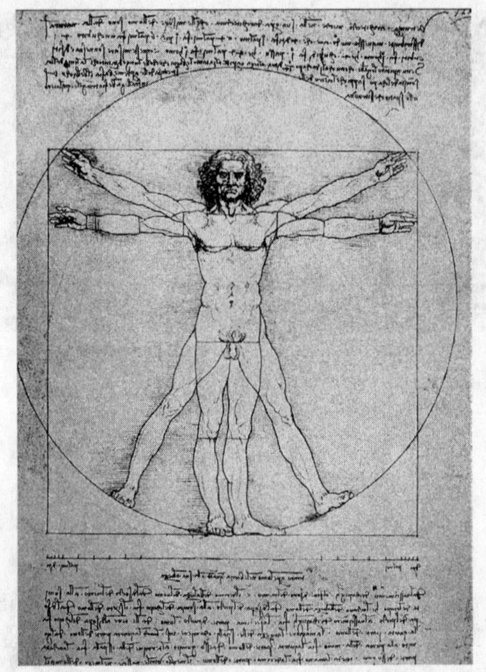

Leonardo da Vinci's "Vitruvian Man."

embodied, standardized measure changes everything. Now the standard does not belong to nature or to community life, but is a special artifact with a unique identity and role. The standard must be specially housed, protected, and maintained. Possession of this standard becomes linked with political and social power, with the authority of kings and the grandeur of God—which is why the Romans kept their standards at the Capitolino, the Greeks in the Acropolis, the Jews in the Temple, the kings and lords in their palaces, the United States near Washington, DC, the French near Paris, and so on. The nation's ruler owned the standards and provided the guarantee of their reliability, while the ruler's aides supervised and maintained them, supplying and inspecting copies. Questions about the accuracy of these copies, how adequately they are kept and how reliably they are used, are then introduced into the use of weights and measures.

When I buy fruit and vegetables at the Union Square farmers' market in New York City, for instance, I trust I'm getting the right amount for my money, even when I do not recognize the sellers, because I know that city inspectors pay regular visits to the vendors to check their weighing equipment. In the Malate district of Manila, a poor working neighborhood in the Philippines, there is even a special term relating to marketplace trust of measures: a *suki* is the local term for someone whose scales you trust because you do regular business with him; when you cannot find a suki, you look around for a friend who has a suki working the market that day.[12] Even scientists do not usually have time to recheck every single material specification when staging experiments; they too have to rely on trust—a trust reinforced by the scientists' knowledge that suppliers have feedback quality-control systems in place that actively look for incorrect measures and instantly change the process if flaws are found, and that suppliers are aware that any slipup means no future orders from those scientists. Weighing and measuring is now a social institution at the center of circles of trust and expertise.

Embodying units also changes the relation of units to each other. In the almanac of the world, the units maintain their own independence and integrity. A hand is a hand, a finger a finger, a seed a seed, and none of these units gains or loses through its belonging to a pattern of relationships with the others. If three rail lifetimes outlast a hound, or fall short, the hound's still a hound. The "ontology" here (meaning existence-character), as the philosophers say, is simple: the primary relationship of these units is to the world; their familial, patterned relationships are secondary and mainly organizational. When units are embodied, however, they can be defined in terms of each other. A foot can be defined as equal to 12 inches, so that this relation is an intrinsic property of each unit. Now the ontology is different: the network rules over the identity of each element in it. The laws of the network govern the units as necessarily as the laws of geometry govern triangles and squares.

The architecture of measuring, too, is changed. In the almanac of the world, we relate to a particular object in measuring—to your foot, a bird's life, a seed or collection of them—and connect it to the world. When measures are embodied, we relate to a network, and it is not really the (replaceable and possibly flawed) specific measuring element that we put into play and connect with the world, but the entire network.

METROLOGY AND METROSOPHY

The embodiment of measures—the manufacture and maintenance of standards, the networks in which they are involved, and the supervision of these networks—gave rise to the subject of *metrology*, the science of weights and measures. Metrology is both a theoretical science, for it involves the knowledge of the network and its interconnections, and a practical science, for it involves the knowledge of how to apply measurement to different domains, from science to economics and education. The study of the cultural or spiritual significance of measures and their patterns—such as the tie between the proportions of the Vitruvian Man and the Greek idea of beauty—can be called *metrosophy*.[13]

Measurement, in every culture, has rich symbolic dimensions. Once measuring is embodied, and metrology becomes a social institution open to issues of trust and distrust, it is no longer a neutral activity but is tied to justice, the good, and human enrichment, with a possible dark side having to do with injustice, exploitation, and alienation. This brings about such changes in the way human beings live that many legends have sprung up about how idyllic life must have been beforehand. Herodotus, the fifth-century BC Greek historian, provided one of the earliest:

> The Carthaginians tell us that they trade with a race of men who live in a part of Libya beyond the Pillars of Herakles. On reaching this country, they unload their goods, arrange them tidily along the beach, and then, returning to their boats, raise a smoke. Seeing the smoke, the natives come down to the beach, place on the ground a certain quantity of gold in exchange for the goods, and go off again to a distance. The Carthaginians then come ashore and take a look at the gold; and if they think it presents a fair price for their wares, they collect it and go away; if, on the other hand, it seems too little, they go back aboard and wait, and the natives come and add to the gold until they are satisfied.[14]

Herodotus's story has been repeated in various versions over the ages, and well-meaning but naïve storytellers have often described supposedly primitive cultures—including Native American and African tribes—as lacking in weights and measures, interpreting this as a sign of their innocence and purity. Such stories are invariably false or fanciful. In everyday life, human beings are continually and routinely measuring the world in myriad ways and for many different purposes, however crudely or informally, and aware of the potential for abuse.

According to some Judeo-Christian legends, Cain is said to have invented weights and measures, so that instead of living "innocently and generously," human beings were now thrust into a state of "cunning craftiness."[15] Measuring was so frequently associated with the potential for crime and sin that the Bible equates accurate measuring with justice itself—whence the traditional image of justice as blind and holding scales, and the commandment to have "just balances, just weights, a just ephah [a unit of dry measure, about a bushel], and a just hin [a unit of liquid measure, about a gallon and a half]." This injunction, in Leviticus 19:36, is among several sacred "Thou shalts" where transgressors faced death as the traditional punishment. In early modern Europe, tensions among local, municipal, and government authorities over who controlled measures for commerce and taxation gave rise to political struggles and sparked riots. Over the centuries, civil and religious authorities attempted to curb abuses of measurement with occasionally severe threats and sanctions, such as chopping off fingers of wrongdoers or instituting capital punishment.

Today, these kinds of abuses are far less common. We routinely purchase food, buy fabric and furniture, and plan to arrive at train stations while relying on measures whose accuracy we do not and cannot know firsthand. How are we sure whether *this* scale is accurate, *that* ruler correct, or *those* clocks right? Yet we do so fairly confidently; the fast pace of modern life would be impossible without it. We do not need complete accuracy, and for most purposes are apt to be satisfied if we think, say, a scale measures meat to about half an ounce per pound (one part in thirty-two), a ruler to an inch per yard (one part in thirty-six), or a clock to a second per minute (one part in sixty). What matters is that we do not have to worry about it, even though we could suffer or get cheated if the measure is off. We tend to trust measurements—and without such trust, modern life would grind to a halt like a machine drained of oil. The deepest paradox of modern life is that we depend so extensively on such accuracy with so much trust. Abuses of measurement in the modern world tend to take a different form: equating the real with the measurable, and putting too much trust in measurement to establish such fundamentally unmeasurable things as intelligence, happiness, self-esteem, educational quality, and so forth.

In relatively recent times a single measurement network, the International System of Units, or SI, has come to be used universally; even those few countries (the United States, Liberia, and Myanmar) for whom the SI is not the official measurement system ultimately define their standards in terms of the SI. The demands on the SI are heavy: today's engineering projects involve myriad interrelated elements which all must be measured alike, sometimes to an accuracy of up to one part in a million or more. Yet the SI's operations, overseen by an international group of scientists who work on diplomatically neutral territory just outside Paris, are all but invisible. Metrologists, I have noticed, like to pass themselves off as colorless people who lead dry careers in a field outside the mainstream of science. Consider this book an exposé of how false that image is. The closer I looked into metrology, the more I found tales as wild, and personalities as outsized and creative, as are found in politics, music, and the arts.

The story of measurement is one of the most spectacular manifestations of globalization. Once upon a time, every region of the planet had its own "noonday cannon"; its own measurement system which had arisen from the local resources and practices to serve local needs. The *local* measurement systems of different societies were as original and varied as their artworks, political systems, and other forms of cultural life, and their views of the point and purpose of measurement equally diverse. The more important a society viewed some aspect of the environment—gold in West African cultures, salt in Mesoamerican communities, court ritual in China, distance in nomadic tribes, agriculture in premodern Europe—the finer and more elaborate measures of this aspect tended to be, and the more these measures were specified and regulated.

Yet within a short time, historically speaking (within about 200 years), virtually all these systems became consolidated into one *universal* system of measurement, adopted by virtually every country on the planet. It is as startling as if the entire world came to speak one language. How did it happen? In Chapters 2, 3, and 4 we follow the story by choosing three systems—length measures in ancient China, gold measures in West Africa, and agricultural measures in agrarian Europe—and watch the different ways they evolved, and how they came to intersect and be transformed. Chapters 5 and 6 trace the march of the metric system as it came to replace the local systems and become universal; Chapters 7 and 8 discuss some of the backlash—serious as well as comic. Chapters 9 and 10 witness the rebirth of the dream of a *natural* and fully unified system of measurement in the form of the SI, while Chapter 11 concerns changes in the meaning of measurement. Chapter 12 is about a comprehensive revision, now under way, of the international measurement structure that underpins global science, technology, and commerce, which has been billed as the system's biggest overhaul since the French Revolution, and which realizes a centuries-old dream by tying units to absolute standards.

The creation of the SI, a unified system, is only one of the radical changes in measurement that we are experiencing. Until recently, the SI—as all measurement networks—had to rely on improvised embodied artifact standards that scientists themselves created: a stick as the basic length unit, a lump of metal as the basic weight. Our global village, so to speak, has been in the same position as the seaside one which marked time with its cannon. In the twenty-first century, this is about to change. Scientists are about to connect all measurements—length, weight, time, and the other fundamental units in the system—to absolute standards far more universal and reliable than even the rising and setting of the sun. Chapter 12 describes this looming epochal transformation of the world's measurement system into something *absolute*: not tied to anything local, or arbitrarily universal, or natural, but to physical constants. No single ruler or state will be privileged to possess the base measurement standards. The standards will not even be in a place. They will be present everywhere in the world around us, available to all with the right instruments. In the epilogue, we note how the meaning of measurement has changed in the modern world.

To find out all of this, we have to put ourselves in the position of the youth in the seaside village.

CHAPTER TWO

ANCIENT CHINA: FEET AND FLUTES

Guangming Qiu waited patiently for me outside the National Institute of Metrology (NIM) in Beijing.[1] It was yet another horribly hot and humid July morning in the city. The temperature had reached 90 degrees by 10 a.m., the air was slightly gritty thanks to dust blowing from the Mongolian plateau about 400 miles to the north, and passersby already looked oppressed by the heat. Not Guangming, a bright and cheery small white-haired woman, now seventy-five, who is the last member of a team of historians that formed at the NIM in 1976, as one of the strange byproducts of the Cultural Revolution. She retired from the NIM 10 years ago but still researches the history of metrology on her own.

She ushered me into a car and we drove to the NIM's new laboratory in Changping, about an hour northwest of Beijing. The Changping campus, which opened in 2009, was built in an area protected by mountains on one side and the famous Ming Tombs preserve—containing the site of the mausoleums of thirteen Ming dynasty emperors—built with ancient Chinese measures, on the other. In that isolated environment, the laboratory can conduct high-precision research into magnetic, electrical, and mechanical standards relatively free of interference stemming from traffic and industry. Its state-of-the-art SI instruments include a brand new "Joule balance," a device whose sole purpose is to evaluate the possibility of replacing the kilogram artifact now stored at the International Bureau of Weights and Measures outside Paris—the world's current embodied weight standard—with an absolute standard. The Joule balance is a novel approach for linking mass with the Planck constant that is not being explored anywhere else but China.

After our tour of the lab, which pointed to the future of Chinese metrology, we returned to Guangming's apartment in central Beijing, to discuss its past. En route, she told me the strange story of how and why she began to research Chinese weights and measures. She was born in Nanjing in 1936. A year later, as the Japanese army marched toward the city spreading murder, rape, and horror in its wake, her parents fled to Chongqing, in Western China, where the Chinese government was relocating. Ten years later, after the war, Guangming's family moved in with a relative living in Hunan province, and her father got a job as editor in chief of the *Hunan Daily*. Guangming studied painting and art at Suzhou College in Jiangsu province. By the time she graduated in 1957, the People's Republic of China (in the midst of its transition to the SI) was assigning people to jobs, and Guangming was ordered to teach painting in Tianjin, near Beijing. Six years later, she was reassigned to an entirely different job—to work in a factory building measuring equipment for the NIM, near its first location in Beijing. She worked there another 9 years.

When the Cultural Revolution erupted in 1972, all research ceased and her factory work ended. I asked what she did during those tumultuous years. "Nothing of importance," she said, quietly, indicating that nothing more would be forthcoming.

BIRTH OF CHINESE METROLOGY

In 1976, in the waning days of the Cultural Revolution, the central Communist Party concocted the idea of a movie about how the first emperor, Qin Shi Huang Di, unified the weights and measures of China. The country's first movie studio, Changchun, began work on a script and wanted to collaborate with the NIM. Institute officials were reluctant to get involved (the turmoil of the Cultural Revolution had not entirely subsided and political interpretations of past emperors were dicey matters) and the project died. But the episode prompted the NIM to establish a group consisting of half a dozen people to research the history of Chinese metrology. During the Cultural Revolution, many researchers had left, so the NIM was shorthanded. To help fill the gap, the director asked Guangming to join. By this stroke of good fortune, she became a researcher in the history of metrology.

Hot sunlight streamed through the windows of Guangming's tiny apartment, and she got up to draw the curtain. She pulled some books off the shelf and opened them to pictures of ancient length measures from Chinese museums. Initially, she told me, the group's research focused on metrology up to and including the time of Qin Shi Huang Di (259–210 BC), the first emperor of unified China. By then, weights and measures already had a long and venerable history in China. Metrology in China was born of the need to satisfy production activities and maintain the state apparatus, coupled with the ancient Chinese passion for "defining and maintaining good order," and appeared as early as the Neolithic era in the third millennium BC. Careful, systematic measurement is evident in the precise manufacture of jade ritual artifacts in the Neolithic era, prior to about 2000 BC.[2] The early linear measures used to make these artifacts were based on body parts, especially fingers and hands; sometimes a distinction was even drawn between measures of a man's hand and of a woman's hand.[3] The principal body-derived

A BRONZE CHI RULER FROM THE ZHOU ERA (1046–256 BC).

measures were the *chi* (pronounced *chyr*, rhymes with "sure"), a foot-measure that could vary from 16 to 24 centimeters, depending on time period and region, and the *cun* (pronounced *tswun*), which was once connected to finger width but at least as early as 400 BC was regulated at one-tenth of the chi. But even as early as the Neolithic era these units were embodied—tied not to the feet of individuals but to easily duplicated measuring sticks.

The diversity of measures persisted from Neolithic times through the Shang dynasty (about 1600–1046 BC) and the Zhou dynasty (1045–256 BC), two long-lasting but not heavily centralized dynasties. In the Shang dynasty, bronze ritual vessels and the designs on them came to be governed by precise mathematical rules. These rules were not simply for the sake of proper proportion or even aesthetics; they were symbolically significant, reflecting deeper proportions in the universe. "Structure was an element in the hierarchy of meaning," the Chinese art history scholar Robert Poor writes in an article on the role of measurement in ancient Chinese ritual, "a metaphor for the moral and spiritual order of the universe made plain for all to see."[4] Meanwhile, bronze bells were increasingly important in Chinese life, thanks to technological developments. From about 1200 BC onward, bell foundries sprang up, especially in the south, to satisfy a steady demand for military music and signaling. The bells' form became standardized, with artisans continually trying to improve their sound.

In the Zhou dynasty, a truly national Chinese culture and state emerged: The written language was developed to a high level; iron use became common; and China's great philosophers, including Confucius, Lao Tsu, and Mencius, delivered their teachings. Throughout this dynasty, as well, the chi was the basic unit for measuring objects of human dimensions; Confucius reports his height as 9.6 chi, and says his father, who stood 10 chi, was about as tall as a human being could get.

Court policies and practices concerning the authoritative rites, too, began to standardize. These included ideas about proper musical performances, and involved specification of the harmonic system.[5] Bells began to work their way into the court ritual-religious system, becoming increasingly important in its emerging musicological component. By 800 BC, court ceremonies of investiture, worship, and liturgies used to mark the ritual calendar and define royal authority came to rely on increasingly elaborate and expensive cast-bronze bells, singly and in arrays. Imperial music masters began to explore note possibilities beyond the three, four, or six bells that were usually available. When a twelve-note system developed, sometime before 400 BC, it further elevated the ritual importance of bell chimes, because their twelve tones were readily integrated into a court mathematical philosophy. "Tuned bell-chimes," says Howard L. Goodman, an American scholar of early China, "could demonstrate to the Emperor's underlings and visitors a totalistic harmony that related to formulas, mathematical completeness, and a mysterious confluence of tunes and numbers, all of which increased the king's importance and ritual power."

Most of China's old capital cities had bell and drum towers that marked off the hours and served as landmarks for city planning. But the ritualized music at imperial courts, which took place in palace rooms or altars just outside the palace, involved a far more intimate interweaving of music into court culture. The ritual twelve-note harmonic-scale system that emerged was called the *lülü*—a deceptively simple name for something as serious and elaborate as a sonic system, consisting of two unique Chinese characters both coincidentally pronounced *lyu*. As archaeologists have discovered, many ancient bell-chime tunings reflected these *lülü* notes. The lowest of the twelve notes was called the *huangzhong* (pronounced like "hwahng-joong"); that term was also often used alone to imply musical correctness. This harmonic system was not equal tempered, with every pair of adjacent notes having identical frequency ratios, as in the Western European classical tradition; it would sound out of tune to our ears and was not the twelve-tone scale, say, of Arnold Schoenberg, though it consisted of a sequence of twelve relatively evenly spaced steps. The names of the individual bells developed slowly over hundreds of years, down to about 400 BC, with different regions using different names. During this time both the nomenclature of the bells corresponding to the twelve lülü steps and the mathematization of the harmonic system became standardized. In 1977, bulldozers leveling a hill to build a factory in Suizhou, China, uncovered the tomb of a minor Zhou-dynasty lord named Marquis Yi, dated sometime after 433 BC, which contained a vast collection of bells complete with musical notations explaining the scale and relationships between keys. Research in the wake of this discovery has vastly improved knowledge of the ancient scale system and its role at court.[6]

Even pitch was important—of "critical importance" at court, writes University of Pittsburgh physicist and music historian Bell Yung, and often exploited "in power struggles among the different factions in the preparation of [a] ritual."[7] To set the base pitch for the zithers, flutes, and singers of imperial court orchestras the twelve steps of the lülü were

DISCOVERED IN 1977 IN SUIZHOU, THE BELL COLLECTION OF ZHOU-DYNASTY LORD MARQUIS YI HELPED SCHOLARS DECIPHER THE ANCIENT CHINESE SCALE SYSTEM.

linked to the dimensions of twelve pitch pipes. These pitch regulators were cast-metal, straight-walled aerophones without finger holes, and their length was specified by court regulations in chi. Thus the chi—the basic unit of length—was inextricably connected with musical pitch, at least at the imperial court. Outside the court little is known and few artifacts exist, though the chi surely was at least loosely governed by the court's definitions.

Nearly every civilization has legends that serve to telescope lengthy historical processes into single episodes. The birth of measures, for instance, is often said to be due to some god or heroic figure: according to a Greek story, Pythagoras invented weights and measures for the Greeks, while an early Roman text says that the god Jupiter gave them to the Romans. The Chinese have corresponding legends. According to one traditional and frequently repeated story, China's first legendary emperor—Huang Di, who lived from about 2697–2597 BC and is often called the father of Chinese civilization—sent a minister to the mountains to find a species of bamboo that was revered for the extraordinary regularity in its length and thickness. He then cut a piece, closed at one end, 3.9 cun in length, and turned it into a flute, whose pitch became the huangzhong. He made eleven more flutes to create the lülü.[8] But this story is certainly as legendary as the ones about Pythagoras and Jupiter, born of the same poetic desire to compress a complex history about an important subject into a single event.

China's first centralized, imperial regime appeared only in 221 BC, when a warlord named Ying Zheng conquered local lords and gave himself the imperial title Qin Shi Huang Di, appropriating the name of the legendary ancestor of Chinese civilization. Qin Shi Huang Di's first act as emperor was to issue an imperial edict unifying weights and measures in the realm, and he ordered the edict carved on or cast into the weights and measures themselves. It was the first unification of weights and measures in China; for a while, Communist Party officials thought this episode had the makings of a revolutionary epic, which led to Guangming's surprising transformation into a historian of metrology.

While the group's research began with the history of Chinese measures up to Qin Shi Huang Di, Guangming told me that after the cancellation of the movie project the research was extended to cover the entire history of Chinese metrology. Group members read everything they could get their hands on about metrology and spent years scouring museums all over China for artifacts. Also, under Mao's regime, beginning in the mid-1950s, archaeology as a modernized scholarly endeavor was highly promoted, and discoveries of Han and pre-Han tombs, artifacts, instruments, and so forth exploded in number. The group had to master and piece together classical texts, the ritual practices of ancient courts, cosmology, music, and the most recent archeological discoveries of metrological and musical artifacts. "Hard work," Guangming said.

The Qin dynasty did not last long and was succeeded by the Han dynasty (206 BC–220 AD), which also issued edicts for weights and measures, and developed scales and other metrological instruments, such as calipers for measuring sizes of round objects. In the middle of the Han dynasty, a high court adviser named Wang Mang, a member of one of the most influential court families, seized power and ruled for a dozen years. "A short but important episode," Guangming said. The historical verdict of Wang Mang is mixed—some view him as a reformer, others as a usurper and tyrant—but "he was good for metrology." He initiated the imperial Chinese tradition of creating and preserving thorough documentation of weights and measures, and the use of cast-bronze metrological devices. From then on, throughout Chinese history for almost 2000 years, the emperor of each new dynasty ordered a review of metrological practices, often resulting in new weights and measures, to make sure that they matched whatever ancient or previous dynasty's example was agreed on by the court's scholars and technicians—and documented the findings.

During the Han era, numerology and mathematics blossomed in court scholarship. This numerology was expounded in scholarly treatises on the court rites, and ultimately it posited numerical relationships among such ritual systems as the lülü harmonic intervals, celestial motions, and the calendar. Having the court instruments pitched properly grew in importance in establishing the legitimacy of the dynastic rituals.

Proper pitch was produced by pitch pipes that had been constructed to classically prescribed lengths. The huangzhong was also involved in the definition of the court's capacity measure. An important court history titled *Hanshu* (whose author died in 90 AD)—literally, *Historical Documents of the Han Dynasty*—defines both the chi and the capacity measure in terms of the number of millet grains placed end-to-end so as to be equivalent to the length of a huangzhong pitch pipe, and the number of grains needed to fill it. Several famous writings around the second to third century BC specify that length as 9 cun, or 0.9 chi. Ninety black millet grains defined the chi, 1200 grains defined the capacity measure, with yet another measure being the weight equivalent of the number of those seeds that fill a huangzhong pitch pipe. In this way, during the Han dynasty, metrology became intimately bound with ritual practices—with the court's religious system, with the symbology of court dress and uniforms, with astronomical observances, and with the lülü musicological system. A change in any one could not be undertaken without exploring the changes that might have to be made in the others.

I asked Guangming if her research included exploring these defini-

HISTORIAN OF CHINESE METROLOGY GUANGMING QIU, WITH CHI RULER MADE OF BLACK MILLET SEEDS EMBEDDED IN A STICK OF WOOD.

tions. "Yes!" she said. An important part of the research done by her history of metrology group was to read the classic documents carefully, carry out their computations and mathematical steps, examine the existing artifacts, and to reconstruct the traditional practices. She left the room to retrieve something from her study. She returned with an armful of wooden sticks, each about a foot long, that she had made in the 1990s at the NIM. Down the length of each she had cut a slot, and in each slot she had pressed a row of "average" millet seeds of the sort that had grown in Han dynasty China, as per the instructions in the *Hanshu*. Metrologists today think that the Han dynasty chi was approximately 23 centimeters, which it took about 90 to 112 seeds to produce. Her sticks showed a chi that indeed fell in that range.

The court metrology—the length and weight standards determined by court scholarship—did not always extend outside the palaces into the towns and countyside. In the marketplace, merchants and craftsmen would not have counted and weighed seeds when buying and selling, but would have used improvised measures. Still, these improvised measures most likely had some loose relation to the court definitions, in the same rough-and-ready way we assume that our marketplace measures bear some at least distant relation to official standards in the vaults of state capitals. Within the imperial precincts, the definitions were an essential part of the court rituals. The ancient Greeks, as we have seen, regarded metrology as having a special spiritual significance when its elements were proportionately related. For the ancient Chinese, metrology also had special significance, but of quite another kind—a social and cultural meaning that was tied up with metaphysical ideas, classical texts, and artifacts. The science and execution of measurement standards in such a ritual context is what inspired Hans Vogel, a German scholar of early China, to apply the term *metrosophy*.

THE POLITICS OF PRECISION

In 274 AD, a court official named Xun Xu tried to introduce a tiny change in the chi. The ultimately unsuccessful attempt and its fallout shed much light on the tight interweaving of metrology, musicology, and politics at the imperial court. This episode is the subject of Goodman's book, *Xun Xu and the Politics of Precision in Third Century AD China.*[9]

Xun was from a politically well-connected family in Luoyang, which had become the Imperial seat of the Wei dynasty in AD 220. He was a junior scholar for the Wei dynasty, one of several warring kingdoms that succeeded the Han. He was also respected as a portraitist and court archivist, and had a keen musical ear. According to an undoubtedly apocryphal story, at some point he realized he needed to re-create a sound he had once heard a cowbell make on a trip he had taken to the north. Later at the capital, he ordered his staff to round up cowbells from that area and to his delight was able to pick out the one he had heard decades before.

In 265 AD, the Wei dynasty was overthrown—and its leader assassinated—by friends of Xun's family. Xun was soon an insider in the new Jin dynasty and a leading member of an ambitious faction that sought to influence the court through its planning of the royal succession—and, in Xun's case, through politically charged technical reforms. He became a prominent archivist and regulator of court institutions and rituals for the Jin. Among his several titles were "Palace Writer" and "Superintendent of the Imperial Library." Around 270, an elder cousin recruited him to reform the new dynasty's musical practices. This was not uncommon: each new emperor would order a scholarly reexamination of the inherited ceremonies to make sure they were technically correct, exercising the virtue of "right conduct" essential to political legitimacy. Most scholars in Xun's position would have made minor changes. Xun, obsessively ritualistic and politically ambitious, immediately introduced significant change in the verses of the ritual songs. Other court officials opposed him, but Xun replied with both aesthetic arguments ("my reforms sound better!") and antiquarian arguments ("this is the way it was in the idealized ancient Zhou times"). His reform was both musicological and political: it implied that the practices of the previous Wei rulers were illegitimate, which cast suspicion on lingering Wei policies and their supporters.

Xun's court duties also included searching palace storerooms for ancient artifacts, which led him to create another court revolution, this one musical and metrological. In 274, he came across a cache of old bronze pitch-pipe regulators called *lüs*, used by previous court musicians to establish pitch for the court instruments.

Most court scholars would have simply used the inherited pitch regulators and found some references in classic texts to justify their use. Not Xun. He compared the sound of the ancient lüs with ones currently used at court, and found the older ones pitched slightly lower. This inspired him to mount a vast project to collect, identify, and compare standards from earlier dynasties. He concluded not only that the current court instruments were literally out of tune with the ancient orchestras and cosmic harmony, but also that the ancient chi itself had become inappropriately long during the last decades of the Han dynasty. "This was not a way to investigate antiquity and honor the sages," wrote Xun, "nor to provide a system for later generations."[10]

Goodman terms Xun's strategy of harking back to the Zhou his "*prisca* Zhou," adapting the phrase from the West's "*prisca theologia*," the belief of early Protestant scholars that adoption of a pure, Biblical form of theology could improve the world and offer its people a single, blessed life, continuous with the past but prepared for an open-ended future. While not quite the same as that in sixteenth- and seventeenth-century Europe, Xun Xu's strategy provided the Chinese court with a similar kind of revisionist critique of the present by appealing to a more distant and authentic past. For example, Goodman says, imagine if Western religious legitimacy "were to have depended on establishing Christ's (or Paul's, or even Gregory I's) exact pitch for psalmody, whether through some vaguely reconstructable pipe, trumpet, or string-length, or, more abstractly, a radical reading of scripture that might point to specifications of magnitudes."[11]

Xun's high status gave him access to workshops and a trained staff. He ordered a new bronze chi standard cast, about 23.1 centimeters according to our measures, about a centimeter shorter than the existing ones made at the end of the Han and imitated in the Wei dynasty. It was not difficult for Xun to revise the length standard himself, in the absence of professional metrologists or independent taxation and mercantile agencies, classes of people whose interests would be threatened by a change. Nor was there any significant commerce with foreign states which might be adversely affected by this change and whose presence would pose an obstacle to a change in measures. Xun's change in the length standard, Goodman writes, was his career signature, "a ritualized search for an ancient, or Zhou, truth."[12] It was as if NIST, the National Institute of Standards and Technology in Washington, DC, and the official metrology agency of the United States, sought to elevate its political profile by insisting that the government use the feet, inches, and pounds that the founders had used in Philadelphia in the late 1700s.

The Jin emperor happily implemented the measure—though in the countryside people familiar with the existing measure resisted. Xun Xu's work reinforced the emperor's legitimacy by showing that the Wei dynasty that the emperor's family had annihilated was ritually incorrect. Xun's corrections also resonated with power struggles, policies, and families jockeying for court control. It should be noted, though, that Xun's resultant new tuning system applied only to the ritual court music and did not affect the folk music known as *yuefu* ("ywe-foo"), which was extremely popular outside court walls after about 100 AD. That music had been especially fashionable in Wei times and was gaining in popularity within court walls. Goodman's research suggests that the *yuefu* style of music at court may have been forced to adapt to the new tuning, and Xun Xu's brilliant maneuver both allowed the popular new music in the court *and* subjected it to imperial control. But, Goodman writes, "Xun's standards did not flow into the practices of palace clothiers, coin casters, and other craftsmen who did not share the need for a *prisca* Zhou reform. The foot-rule as a metrological 'overseer' did however remain a mainstay of court rites and even court liturgies."[13]

Xun Xu went even further in musicological reforms. Based on his new chi magnitude, Xun made twelve pitch-pipe tone regulators, which in turn were used to correct the court orchestras' flutes, known as *di* flutes (pronounced "dee"). These instruments in various forms dated back many hundreds of years. Open-ended and made of bamboo, the flutes played pentatonic or heptatonic scales. Using a remarkable algorithmic-like process, Xun adjusted the spacing of the finger holes according to the new chi and its derivative pitch regulators. That process, though it didn't yet solve the problem of equal temperament, sought to impose the ideal harmonic system of the lülü upon the problems of tone and pitch in real instruments.

In the 1980s and 1990s, Xun Xu's flute finger-hole spacing system was extensively studied by the Chinese music archaeologist Zichu Wang, who used a Stroboconn, an electronic device to measure musical pitch, to compare sounds produced by cylinders constructed with equal hole-spacings with cylinders constructed according to Xun's algorithm and patterned after the design of ancient flutes.[14] Wang was particularly interested in exploring the "end correction," that is, whether Xun com-

PAINTED POTTERY FIGURINE OF A MAN PLAYING A DI FLUTE,
FOUND IN AN EXCAVATED TOMB IN THE SZECHUAN AREA,
FROM ABOUT AD 220–265.

pensated for the fact that the structure of a fundamental pitch is changed by the pressure wave of the escaping sound wave from the finger holes. In 2008, Goodman, who early in his career studied music at the Juilliard School, teamed up with Y. Edmund Lien, a former engineer who had changed careers and became a graduate student in Chinese literature at the University of Washington, to analyze Wang's work. They concluded that in the third century Xun had not yet grasped the physical phenomenon of end correction (though some Chinese scholars dispute that); the effect would remain unaddressed by music scholars until the Muslim philosopher-scientist Abu Nasr al-Fārābī tackled it in the tenth century. Nor was Xun seeking equal temperament. Rather, he was seeking through his algorithm how best to accommodate the numerological lülü scheme to the demands of real performance and the need to use the chromatic scale to play in different keys. For the first time, Goodman and Lien write, Chinese court orchestras had "flutes that used the ritual pitch-standard pitches so as to be in tune when responding to mode and key variation." Through his meticulous attention to detail, Xun in a certain sense brought the court music into "real-world physics and acoustics."[15]

Xun's metrological reformations did not last. His strident faction ran afoul of court politics, and he himself was accused of bad scholarship and defective aesthetics—he made the flutes play too high, charged critics. He was forced out of office, transferred to another with no metrological or musicological power. Archeological evidence shows that chi standards in various contexts grew longer again within only a generation or so after Xun's death. Though Xun's episode resulted in only a momentary change in the length of the court chi, the volatility of the length measure and its dependence on musicological issues, as deciphered by the various research of Goodman, Vogel, Wang, Lien, and others, reveals much about what was strongly coupled to what at court in the third century AD—and in fact for long stretches of Chinese imperial history.

This connection between musicology, metrology, and court politics persisted for over a thousand years. China was isolated from foreign pressures that would lead it to change a system that worked. During the Ming dynasty (1364–1644), the legendary admiral Zheng He took an armada of hundreds of ships and tens of thousands of crewmen throughout Southeast Asia and the Indian Ocean, seeking to expand Chinese influence and control international trade. Disappointingly, Zheng He encountered only civilizations with little more to offer than raw materials, and so the expeditions were discontinued "for the same reason the United States stopped sending men to the Moon," says political scientist Jack Goldstone: "there was nothing there to justify the costs of such voyages."[16] What little trade resulted from these voyages inaugurated no challenges to the long-standing Chinese metrological traditions. The first serious disruptions would happen only hundreds of years later, at the end of the Qing dynasty (1644–1911), in the wake of the Opium Wars.

Guangming's group published numerous books and articles starting in 1981 and continuing into the 1990s, documenting the history of Chinese weights and measures from early times up to the introduction of the metric system in the twentieth century. In 1992, they published a comprehensive survey.[17] One by one, the members of the NIM's history of metrology group quit or retired, and were not replaced. "Hard work, low pay," Guangming says. "Others in the Institute looked down on us, because we were not in the natural sciences, and did our research with documents and old things. Nowadays, people want to work in the lab, with high technology and modern equipment—what we saw at Changping—and are not interested in the history." Soon she was the only one left to research the roots of Chinese metrology. When she retired in 1999, no one replaced her, but she continued to research and publish on ancient metrology on her own.

Why did she keep going? "I don't want a luxury life," she said. "I have enough food, enough clothes—my one desire is to do something the right way." Being a scholar of ancient Chinese metrology—researching its ingenious artifacts, colorful individuals, and compelling tales—was as exciting and rewarding a career as she could imagine.

CHAPTER THREE
WEST AFRICA: GOLD WEIGHTS

China's ancient and unique measurement practices persisted so long because of the country's extreme isolation and also the advanced stage of China's development compared to any other culture within reach. A third of the way around the world, in West Africa, a strikingly different measurement practice arose that persisted for the opposite reason: because it could easily coexist alongside the practices of foreign traders from several lands who were always dropping by.

NIANGORAN-BOUAH: GOLD WEIGHTS AS AKAN ENCYCLOPEDIA

In 1959, Georges Niangoran-Bouah (1935–2002), then a West African student, was working on dissertation research at the Black Africa department at the Musée de l'Homme in Paris. It was a heady time to be a young African intellectual. Independence movements were cresting. Ghana had just become the first sub-Sahara nation to achieve liberation and a dozen others would soon follow, including Niangoran-Bouah's native Côte d'Ivoire (Ivory Coast). Imagine the young scholar's consternation when the head of the Black Africa department asked him to compile information sheets on the museum's collection of brass weights—something not on what an Africanist would consider a list of hot topics.[1] Also, the weights were obsolete; the country was using Great Britain's imperial system of weights and measures.

The Akan ethnic group, which is spread across Ghana, Togo, and the Ivory Coast, used these brass castings to portion out small amounts of gold. Akans, whose presence in West Africa goes back at least 2000 years, had developed a thriving trade economy using gold dust as currency by about the fourteenth century AD. They obtained gold by panning it from riverbeds and coastal areas, and later by mining. In shops and marketplaces, they carefully measured out the dust by using scales and other accessories, which is why Westerners started calling the region the "Gold Coast." The earliest African weights were of seeds, stone, and pottery, but when caravans of Islamic traders began coming through in the Middle Ages, Akans began imitating their practice of using cast-metal weights with Islamic-style simple cubic or conic designs. The first Europeans, who arrived in 1471—first the Portuguese, later the British, French, and Dutch—brought other kinds of weights. The Akans' decorations grew increasingly varied and rich, and came to include not only intricate designs and polyhedra, but also flora and fauna, animals, people, tools, sandals, and furniture. Almost everything was fair game, except for a few animals such as cats, owls, and vultures, which were exempted from representation for religious or mythological reasons. Every head of an Akan household had a bag—*futuo*—of such weights and collection of implements to use in weighing gold, and nearly every element of Akan life appeared in some form or other on this miniature theater. Each casting was unique, and the different design, figurine, or ideogram bore no connection to how heavy the casting was. From the 1400s to about 1900, millions of these brass castings were made in West Africa. Hundreds—thousands, in some cases—filled drawers of museums in the West. Those with more-recent Western-inspired designs such as cannons, guns, boats, and locks and keys were particularly embarrassing to purist Africanists.

An ambitious young scholar, Niangoran-Bouah viewed the castings as trinkets, barely more significant than charms from a bracelet. He literally cursed the professor who assigned him this apparently menial chore. Everything else about African culture seemed more relevant to the radical changes underway: more connected with Africa's esoteric knowledge and spiritual powers, with its special laws and emergent philosophies, and with its difference from the West. But he resigned himself to studying the objects. He read what European ethnographers had written and followed their methods: he formulated hypotheses, drew up questionnaires, and set out for the field to question informants—his countrymen.

Their reaction baffled him. Poor, and with no incentive to participate, they tended to answer pragmatically, in the way most likely to elicit cash. More disturbingly, they claimed not to understand what Niangoran-Bouah was talking about. He was outraged; they were treating him the way they would a white tourist. He wrote later, channeling his younger self, "What right had illiterate individuals to contest a study which required several years of research in Europe's most famous library and museums?" Slowly it dawned on him that he was *acting* like a white tourist, and that he had to shed his inherited European assumptions.[2]

Learning the local languages helped. Europeans translated the Akan term *yôbwê* as "weight." Then Niangoran-Bouah listened. "A weight," he wrote, is the European word for "an element in a measuring system which is used to determine the mass value of another object: it can only be conceived within a measurement system and as a rule it can have no other function." But, he continued, "*Yôbwê* (pebble or stone) has a wider meaning; it can be used as a measure and may also have other functions." When it does, it tends to be called other names.

A *dja-yôbwê* (dja pebble), for instance, was a stone whose content was linked to the Akan cultural heritage; a *sika-yôbwê* (gold or silver pebble) was a sum of money; an *ahindra-yôbwê* (proverb pebble) meant a stone bearing a thought; a *nsangan-yôbwê* (fine pebble) was a stone used when paying a fine or tax; a *ngwa-yôbwê* (gaming pebble) bore signs used in games and interpreting hidden meanings; and so forth.[3] All were treated by Western scholars as weights. Literally speaking, Niangoran-Bouah concluded, none were weights in the Western sense at all. While Western traders viewed weighing as an interaction in which one uses an object in conjunction with an instrument to attach a numerical value to one specific property of another object, the Akans used the brass castings together with a set of other apparatus in a complex social interaction. These castings were more like standards for pricing; their weight stands for an amount of gold dust to be exchanged for taxes, fines, services, goods, or other things:

> For the Western-educated scientific researcher, the idea of a balance, and metal mass placed in the scales of a balance, calls to mind weighing, or determining the mass value by comparison with another metal mass. For the Akan natives, balance, pebble, spoon and gold dust all refer to the action of evaluating money. Consequently, there could only be misunderstandings and "deaf and dumb dialogues" between Europe and Africa.[4]

Furthermore, there was no correlation between the designs of the weights and their value, and no absolute weight value existed. European researchers had tried in vain to discover correlations between the weights and some natural standard—seeds or berries, for instance.

"Akan merchants handling gold were not necessarily interested in determining the exact weight," the scholar of Africa Albert Ott warned. "The scientist who attempts to establish an accurate system from existing gold weights will fail. He can never hope to obtain a set of weights which corresponds exactly to arithmetical equivalents."[5]

Why then, Niangoran-Bouah asked himself, didn't the Akans—great imitators!—copy the more precise European system of money and weighing when they saw it? It would have been much easier to manufacture Western-style coins and weights, which were identical tokens, than for Akan craftsmen to hand-make each Akan casting with a novel design. Furthermore, Western coins are easier to carry around than the castings, which were kept in the futuo with accessories: scales, sifters, pans, spoons, and feathers for cleaning, parts of an elaborate and socially complex weighing system. Finally, the Western system was easy to explain and mastered by children, while the Akan gold dust currency system was so complex that instructions on its use were part of adult initiation rituals.

The resistance to change, Niangoran-Bouah decided, revealed how fundamental the brass castings were to Akan culture. The castings he had initially despised as curiosities were keys to the fundamental workings of Akan society and its institutions. The castings, as well as the ritual masks and other objects proudly exhibited in museums, were a valuable window into the lifeblood of Akan culture—its law, finance, commerce, education, mathematics, philosophy, literature, leisure, religion, and mythology—and its difference from the West. In the three-volume book that grew out of Niangoran-Bouah's dissertation, he provided examples for the benefit of his Western readers of how the system operated in practice and composed a "play" whose plot consists of an Akan monetary transaction in which someone borrows and then repays an amount of gold, in which Niangoran-Bouah was able to

Contents of a futuo.

exhibit the social context, the course of the discussion, the role of the weighing, and so forth.[6]

Niangoran-Bouah's dissertation includes a dramatic—though possibly apocryphal—tale of the use of the figurines the first time that European sailors saw Akan gold. The incident, he wrote, took place in the Issia region of West Africa in 1471, when Portuguese sailors interested in establishing trade noticed a chieftain wearing gold ornaments. One sailor pleaded with him for some gold, offering his pistol in return. The chief refused. The sailor persisted, begging, haranguing, and cajoling him without letting up. The chief's advisers laughed at the antics of this strange white man and told the chief to refuse. The chieftain thought for a while. Then he pulled out a bronze box containing gold powder and some figurines, as well as a scale and a spoon. To measure out the gold he chose a figurine called Odiaka ("If he eats, it stays"), a crocodile with a fish in its mouth. When the chief put Odiaka on the scale's pan, the advisers suddenly hushed. As the chieftain carefully spooned gold dust onto the pan, his advisers continued to look on in stunned silence, while the white men excitedly and happily exchanged written notes. When the chieftain finished, he wrapped the gold in cloth and handed it to the sailor.

Niangoran-Bouah explained that the crocodile was a symbol for a powerful person, the man with the upper hand. The king's comrades grasped that he was telling them that the Portuguese sailors were a deadly threat. "If he eats, it stays" meant that, once the crocodile grasps what it desires, nothing in the world can free it again—it's as good as lost. The chieftain was telling his people by the mute act of choosing the figurine, that if he refused the armed white men, they would take what they wanted anyway. The chieftain was saying that he had no option but to proceed with the trade in as dignified a manner as possible. Niangoran-Bouah remarks:

> The European, obsessed with acquiring gold, had no idea that the old Issinian weighing his gold with scales hooked by a string to his left thumb and using original weights, was speaking with his ethnic brothers. The noisy laughter of the Africans who witnessed the scene and the silence which followed it were in contrast with the lack of reaction among the whites. This difference, and the profound gulf which separated them, was underlined in striking fashion by their behavior. Those who witnessed the scene represented different races, different civilizations and different worlds. These people, who needed to trade with each other, did not understand each other. . . . With this scene, we are in the presence of two different methods of expression: the white man's of expression writing—the word, as represented by graphic signs and marks on a flat surface, and the African use of the material image.[7]

Niangoran-Bouah has been criticized by scholars for exaggerating the symbolism of the castings and overstating their cultural role, possibly a reaction to centuries of colonizing powers who dismissed local Akan customs as barbaric, even uncivilized. His claims that the Akan system of weights and measures is separate but equal to the Western one; that the weights contain "the sum total of [Akan] knowledge" and amount to "an encyclopedia of another sort from another world"; and that they are the Akan equivalent of the Bible tend to be regarded today as over-dramatizations.[8] Later in his career, Niangoran-Bouah was accused of further excesses when he claimed that West African drumming was a language on a par with other languages and could be studied by a new science that he called "drummology."

Nevertheless, his work helped open up a new appreciation for the workings of West African culture among African scholars, and he had been set on the path in the least expected of ways: by how its people measure. When the Nobel prize–winning writer V. S. Naipaul visited West Africa in 1982, he visited the now-famous Africanist, by then a professor of anthropology at Abidjan University and, in a profile that wound up in the pages of *The New Yorker*, noticed a collection of gold castings still adorning his desk.

TOM PHILLIPS: GOLD WEIGHTS AS AKAN SCULPTURES

British painter and sculptor Tom Phillips appreciates Akan gold weights from an entirely different angle. One day in 1970 Phillips, a young artist who owned a few African masks made of wood, noticed a weight for sale in a London gallery. "It was a fish—a catfish—curving around on itself. I bought it for my daughter," Phillips told me in a conversation of short, enthusiastic sentences. "She never got it. I looked at it again: Why was it a fish? Why that fish? What was it for? How beautifully it was made!" He bought another, and another. He visited

West Africa to look for more of the weights, and now owns some 4000, which he occasionally puts on exhibit. Phillips estimates that in his hunts he has handled a million of them—a substantial fraction of the total number produced over six centuries—while rummaging through the drawers of London dealers and African shops and marketplaces. He calls his 188-paged art book, *African Goldweights: Miniature Sculptures from Ghana 1400–1900*, a "praise song."[9]

I met Phillips at a restaurant in New York City, not far from a gallery that was about to open an exhibition of his lively and inventive paintings, sculptures, and mixed media work.[10] To my eye, at least, except for one sculpture of assorted figures on a gameboard-like surface, his work had little obvious kinship with the Akan weights despite the time and energy he has invested in them. I asked him if they had influenced his art.

Phillips emitted the sort of kindly but pained snort that lets you know you are barking up the wrong tree. "Everything influences your art," he said. "Of course there's common ground. What pleases me in life is what draws my attention to these weights."

What's that common ground? "Their makers are in love with life. They're fun-loving. They weren't in it to write scripture. Or to create a Dan Brownish secret code. They just asked themselves, 'What new thing can it give us pleasure to represent?'" The Akan craftsmen, he said, took a utilitarian excuse to celebrate the world around them. Over six centuries this celebratory impulse, through their imaginations and virtuosity, took in most features of Akan life. As Phillips writes in *African Goldweights* (where he allows himself somewhat longer sentences):

> With no apparent plan or original intention, they eventually constructed a comprehensive three dimensional encyclopaedia of the fabric of the society, its goods and actions, its characters and roles, its wealth of animals and birds; all things natural and made, with a scope of rep-

> resentation from the grandest chief in his pomp to the humblest tool for working the earth. The product of this haphazard and piecemeal feat was, by virtue of usage, not concentrated in a single place but dispersed throughout the region in the bags of weights that by the 19th century virtually every family possessed."[11]

Many Europeans, he continued—even many Africans—yearn to find some metaphysical or mystical order that the Akans had encoded in their weights, but it just isn't there. Enthusiasts of the weights, he told me, also often sought unsuccessfully to find a link between design and weight value: "That ate up thousands of hours of frustrated enterprise!" It is clear, he said, that the weight value of the castings was not tied to a simple numerical or mathematical product. "The weight value arose from bargaining a series of agreements, augmentations, and approximations between highly disparate objects that themselves often related to different weight traditions and scales from Islam and Europe."

I found Phillips to be utterly unapologetic about the delight he receives gazing at the objects with "the ever more unfashionable eye of a European connoisseur." Yet he is not a total outsider, for his experiences as a practicing artist and art appreciator give him a kinship with the Akan artisans across the admittedly yawning cultural divide. It is impossible to believe, he writes, "that objects of this quality and refinement were made without pride on the part of the goldsmith, and occasional delight on the part of his client. Such pride and delight could only be based on aesthetic response and artistic discernment even if no native terms of evaluation have been recorded."[12]

Phillips's own knowledge as a practicing artist is revealed in passages describing how elaborately planned, carefully constructed, and vigilantly cast the objects had to be. The method is known as the "lost wax" process. The models were thought out from the beginning, every detail separately constructed and then fitted together. When the figure was a human, twenty separate elements can be involved: "arms, legs, head, neck, even buttocks, nipples and navel were fitted to a torso with each joint carefully made firm both with finger pressure and hot metal needle." The inexperienced artist, Phillips writes, is "Gulliver in Lilliput with fingers and thumbs suddenly seeming enormous as one tries to manipulate small threads and beads of soft wax, positioning them to stay where one wants."[13] In the equatorial climate, you have to work quickly lest thin parts droop before completion. As if constructing the wax figure were not enough, casting is another elaborate process. The figure is carefully packed in clay to make a mold, with tiny wax threads called sprues attached to the model to leave passageways through which the wax can be poured out and metal poured in. The mold is heated, the metal liquefied, the now-liquid wax emptied out, and the molten metal inserted. "It is a long drawn out business," Phillips writes. "Most of the time spent is waiting for this to dry, that to cool or this to heat up and liquefy. The timing of each episode, using, as it were, an oven without dials or controls, has to be calculated from experience."[14]

Phillips has a knack for being able to describe the wit and visual appeal of the castings without the mind-numbing and judgmental jargon of the art critic, or the metaphysical moonshine of many an armchair academic. Here's one of his captions: "Two grasshoppers copulating, cast in flagrante." Commenting on an illustration of a damaged figurine of a naked warrior sitting on a stool: "Had his legs been as thick as his penis they might also have survived intact."

Over dinner, Phillips expressed his opinion that the entire history of sculpture and its possibilities were played out in the history of the weights. "It's a miniature art form which you are making in great numbers. You get to try all sorts of things! If your career is making large sculptures, you don't make many objects. You don't have the time or

SELF-REFERENTIAL AKAN WEIGHT,
PORTRAYING A WEIGHING RITUAL.

resources! If you make objects the size of a sugar cube, you can make a lot of them. You try a lot of different ways. Over six centuries, you can try an enormous number of different ways!"

I asked Phillips about his claim in the book that the weights both inform modern art and are better understood thanks to it. "Take an obvious example: minimalist sculpture," he said. "Quite unlike nineteenth-century sculpture! The Akan weights which are akin to minimalist sculpture in appearance must have looked like nothing to the Victorians. They had no point of reference to associate them with art. Nothing! Maybe some ornaments that the Victorians never categorized as art anyway. It's only as art itself progresses that then we Europeans can read back aesthetic quality in those old things. Be able to say, for instance, that one is like a Judd or a Brancusi."

Phillips opens his book with a description of what he called a "grail" object, a remarkably self-referring weight that he had searched after for years, one portraying a group of figures about to conduct the ritual of weighing gold. The figures are intently focused, facing one another across scales and weights.

One of them holds the balance while his fellow trader carries a small container for his gold dust in one hand and a scoop in the other. A third figure ruminatively smoking an African-style pipe seems somehow to be acting as umpire. Although the whole event occupies a space only 3 centimetres high and 5 centimetres long it conveys not only the delicacy of the scales with its two pans suspended on threads but features, on the floor between the protagonists, a couple of tiny shapes, one no more than a millimeter in diameter, that are unmistakably intended to represent weights of traditional form.[15]

Shortly after beginning to collect the weights, Phillips sought out Timothy Garrard (1943–2007), the principal authority on them. Garrard was a British lawyer who had gone to Ghana in 1967 as the only European working for the Ghanaian attorney general in the capital and was soon Ghana's senior state attorney. After his arrival, Garrard became enchanted with the weights, and began collecting them and investigating their history. He thoroughly prepared himself for the task by apprenticing himself in the lost-wax method to an Akan craftsman, earning an MA in archaeology at Legon University—and later a doctorate at UCLA—and in 1983 becoming a museum curator in addition to his legal duties. Garrard became well known and trusted throughout West Africa as both a lawyer and scholar—one of the few, and perhaps the only, European to become a member of an African secret society, the Poro society of the Senufo people—and assembled an extensive collection not only of weights but also of history and lore regarding their use. His book *Akan Weights and the Gold Trade* is the single most authoritative source on the subject.[16]

When Phillips met Garrard, he discovered him to be "a character right out of a Conrad novel—not the sinister Kurtz kind but the engaging, eccentric, born-in-Europe-but-made-in-Africa kind." Garrard invited Phillips along on journeys to track down weights and to track down reports by travelers of their use. Here is an especially vivid and accurate description that they turned up, written by a Swiss missionary in the late nineteenth century:

> The weights, spoons and gold pans are carried with the scales in a leather bag, without which the rich never venture out; they are carried in front of him on the head of a slave. . . . The gold is weighed out amidst very noisy scenes. Each person carries his weights with him, but those of the vendors would be found too heavy and those of the purchasers too light. Arguments go on for a long time until finally the correct weight is produced. Only now does the weighing take place, and the pan with the gold should sink just a fraction lower than the pan with the weights. New arguments break out as the examination of the gold follows. Each grain is turned over: "This is bad gold, look, here is a stone, that piece must be exchanged." There is another weighing and more arguments until after many delays the small transaction is concluded.
>
> The weighing of gold can be as delightful as it is tiring, especially in the case of small purchases, for example when you want to buy fruit or vegetables from the market-women. The law of Kumasi forbids any woman to hold the scales in her hand. How suspiciously she watches you, how sharply she criticizes your weights, and then she begins to argue that the pan with the gold has not sunk deep enough. Finally, when this obstacle too is overcome, she takes the small heap of gold-dust in her hand and separates it into two parts with a mussel shell. One of these parts, she declares, is bad gold and must be exchanged.
>
> In this way a transaction worth a few pennies takes up as much time as one involving several ounces. This can teach a white man patience,

but in my case I lost patience again and again and often abandoned a purchase that was dragging on interminably.[17]

Adding to the time and frustration was the fact that both buyer and seller often weighed the same amount of gold, with their own weights. Here is another description of a weighing, from a British journalist and traveler in the mid-nineteenth century:

> A ha'porth of bananas can be purchased with the precious metal, as I have seen myself, a few grains being placed like a dose of morphia on the point of a knife, and received in a little cloth rag. It therefore results that all grown-up men and women are gold-mongers, and have tests of their own which they correctly and rapidly apply.
>
> The gold being handed over to Amoo [a gold-taker], he took it a little at a time, placed it on a blow-pan, and adroitly puffed away at the base dust of earth with which it had been mingled. The nuggets he cut in half with a knife, rubbed them on a touchstone, and carefully examined the color of the streaks. The gold having thus been taken or assayed, he proceeded to weigh it, partly with little red berries and partly with native weights cast into forms of beasts and birds.[18]

Over two decades, Phillips and Garrard made several trips throughout Africa looking for weights and informants. On one, they encountered what may well have been the last person alive with firsthand knowledge of how to use the weights, a frail, nearly blind man in his late nineties who was delighted to demonstrate the proper way of using the tools for cleaning and dusting, and of handling the scales—positioning your fingers in a way to show that no cheating was going on—that he had learned as a youth at the end of the nineteenth century.

A civil war in the Ivory Coast forced Garrard to flee Africa suddenly, leaving behind proofs of a new book. He fell ill from dementia, and between the war and his health was forced to give up his career and collecting and return to Britain. Garrard sold his house—a mud hut with pointed thatch, which had been built by Akan villagers—to Phillips, and gave Phillips his collection of weights, largely because he wanted it to be in safe hands, not in a country often torn by civil war. He and Phillips made plans to work together on a book, but Garrard grew too ill for the project; Phillips dedicates his own book to Garrard as a memorial to his friend's achievements.

When Garrard died in 2007, Phillips composed an obituary for his mentor which cited one of Garrard's favorite goldweight proverbs, about pioneers and those who follow them: "One who follows in the tracks of an elephant does not need to get wet from the dew on surrounding bushes."[19] Phillips concluded, "His work has beaten paths through virtually unexplored terrain leaving many grateful scholars with information that would otherwise have been lost"—information about one of the most original, innovative, and social measuring systems ever devised on the planet.

CHAPTER FOUR

FRANCE: "REALITIES OF LIFE AND LABOR"

Brigitte-Marie Le Brigand gamely set her shoulder against the gray steel door—it was about twice her size—and gave a heave.[1] Screeching and balking, the door slowly crept open, uncovering a huge metal safe. Le Brigand, an archivist at the French National Archives, plucked an old-fashioned, 6-inch-long key from a 200-year-old wooden box, unlocked the safe, and swung open its doors. Inside were dozens of red archive boxes containing the most important historical documents of the French government.

We reached this miniature fortress by walking through a maze of rooms in the National Archives, located in a palatial estate a few blocks from the Louvre. The building was erected by Napoleon III in 1866, who wished to house the nation's archives in a building whose dignity was in line with the treasures it would contain. The Archives' central chamber would have been the king's bedroom, if the building had functioned as a palace. The walls of the room, floor to 20-foot ceiling, were crammed with shelves of documents, upper ones accessible only via a ladder that clung to a narrow balcony. Display cases were filled with national seals and medals, some dating to the twelfth century. Dead center against the far wall was the safe Le Brigand had just opened. When first constructed, during the dangerous times of the French Revolution, the safe had three locks opened by three different keys in the possession of three different people: the president, the chief archivist, and the secretary of the assembly. Today, Le Brigand said, the archivists only bother to use one lock.

Le Brigand pointed to some boxes. "That contains the papers of Marie Antoinette, those are the papers of Louis XVI, and there is the current French constitution." In the middle of the safe was a shelf with a collection of strange-looking objects. One was a folded-over mass of yellowing paper. "That's the first version of the Declaration of the Rights of Man, written in 1789," she said, referring to the primary human rights document of the French Revolution. It was written by the members of the National Assembly, the first of a series of legislative bodies that governed France during the French Revolution. Article 1 read, "Men are born and remain free and equal in rights," and the document laid out rights deemed to be universal for all human beings at all times. Next to it on the shelf was a mass of twisted copper sheets about the size of a large book. "That's the Constitution of 1791, the first French constitution." Written by the National Assembly when the king was still alive and the Revolution's radicalism had not yet crested, this document created a constitutional monarchy, giving sovereignty to the people. "When it was created," Le Brigand said, "it was given a special copper cover to signify its importance. After the king was beheaded in 1793, and a new Constitution written, the revolutionaries thought they should symbolically break the first one." They had clearly enjoyed doing so; the sheets of copper looked like they had been hit by sledgehammers.

Nearby were two boxes, a black and octagonal one the size of a jewelry box, and a long and thin brown one. Those contained what I had come to see. Le Brigand put the boxes carefully on a table, pulled on a pair of rubber gloves, and opened the boxes. In the long and thin box was a metal stick an inch or so wide and just over a yard long. Le Brigand picked it up and turned it over: no markings. Inside the other box was a simple cylinder of metal an inch and a half in diameter and just as high. No markings, either.

Le Brigand said, "These are the two original *étalons* [standards] of the meter and the kilogram. They were made by order of the National Convention"—another legislative body among those that had succeeded the National Assembly—"and presented to the government in 1799. The meter was a fraction of the earth's meridian, and the kilogram

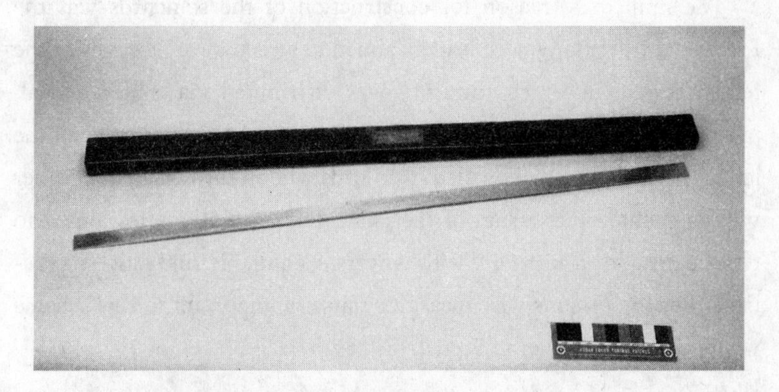

The Meter (above) and Kilogram of the Archives (the cylinder below at left), with accessory measures.

the weight of a cubic decimeter of water. They were meant to embody natural standards, unchanging natural phenomena."

I stared, awestruck. Despite their lack of identifying marks, nothing like these two objects had ever been made before. They were powerful things: for almost 100 years, they had governed a network of measures that had reached over the globe. They were both objects and institutions. They represented the first attempt—unsuccessful, it would turn out—at a final solution to the noonday cannon problem, or finding a way of tying measures to natural phenomena so that if standards were lost, they could be reconstituted with identical measures.

The immediate reason for construction of the standards was the French Revolution, whose leaders aimed to sweep away vestiges of the feudal system in which authority was distributed via a hierarchical, pyramidal arrangement where the king (who technically owned all the land) was at the apex and ruled over lords who in turn ruled over other vassals granted possession of the land. The revolutionaries aimed to replace this feudal authority with universal, equitable, and rational practices. But the reasons why measurement was important to this scheme went back much earlier in French history.

EARLY FRENCH MEASURES

The history of early European measures is tangled, and controversies persist about the origins even of basic units. European linear measures, writes art historian Peter Kidson, are renowned for being a "treacherous quicksand of speculation and controversy from which the prudent are well advised to keep away"; Kidson considers those who do not as a "lunatic fringe which sees order and continuity where everyone else sees only chaos and confusion."[2]

By the beginning of medieval times, about the fifth century, measures in France and elsewhere in Europe already reflected a ragout of influences. Gallic tribes had their own weights and measures, but after Caesar's Gallic Wars the Roman Empire succeeded in introducing many of its basic units—the *pes*, or foot, for length; *libra*, or pound, for weight[3]—in France and throughout Europe, and these units replaced or modified what had been used by the conquered tribes. In France, the foot measure was called a *pied*, which was divided into 12 *pouces* (inches); 6 pieds were equal to a *toise* (fathom). The basic weight measure was the *livre* or pound. Elsewhere in Europe, the German *Pfund*, Dutch *pond*, and British pound also stemmed from the Roman pound. Certain pre-Roman measures persisted into the Middle Ages, including some measures of farmland and distance, such as the arpent and the league. The use of numerals in counting units reflected Arabic influence. Each country, and even different regions within a country, adapted Roman units in its own way, altering the dimensions and the names of measures according to local needs and conditions, and even to suit the things measured.

Various rulers attempted to impose a single consistent measure on their territories. In France in 789 AD, Charlemagne was the first to do so, and put into use standards sent to him by the Arabic Caliph Hārūn al-Rashīd; after Charlemagne's death in 814 the reforms did not long survive. Legend had it that the inner gate of the entrance to the Louvre, constructed in the twelfth century, was exactly 12 pieds—2 toises—wide, and that dimension was held by tradition to provide the size of the toise for centuries. King John (1350–1364) had length and weight standards constructed, now preserved in the Conservatoire National des Arts et Métiers in Paris. In the same corridor of that museum is also the so-called Pile de Charlemagne, or weights of Charlemagne, which were made in the fifteenth century and named after what were said to be Charlemagne's standards. Its pound was divided into 2 marks, with each mark divided into 8 ounces, each ounce into 8 gros, each gros into 3 deniers, and each denier into 24 grains—a grain often said to have been originally a wheat kernel.

PILE DE CHARLEMAGNE, WEIGHTS MADE IN FIFTEENTH-CENTURY FRANCE AND NAMED AFTER CHARLEMAGNE'S OWN STANDARDS.

These royal measures may appear to have been organized and uniform, but those in the countryside were not. As the French metrologist Henri Moreau wrote:

> The units varied, not only from country to country, and sometimes (as in France) from province to province, but even from city to city, and also according to corporation or guild. Of course, this state of affairs led to errors, frauds, and continual misunderstandings and disputes, to say nothing of the serious repercussions such a situation was bound to have on the progress of science. The multiplicity of names given to poorly determined units and the diversity in the multiples and submultiples of the principal measures increased the confusion.[4]

Measures, after all, are tools; people use them for particular ends, and if conditions change or new ends arise measures are adapted or replacements improvised. But measures must be shared and trusted by communities. As a result, they acquire lives of their own, spreading slowly and replaced reluctantly. An interplay develops between tradition, how one measured in the past, and evolving needs. China's centralized society, and relative isolation from foreign influx and trade, kept its needs relatively uniform and stabilized its weights and measures. In West Africa, traders and merchants managed to use theirs alongside those of foreigners in a kind of bemused and benign coexistence. France had a completely different social and economic structure, one which proliferated rather than consolidated measures. Its diversified and changing workplace environments, from farms to craft guilds and commercial contact with other regions in Europe from Norway to southern Spain, each of them with different kinds of workplace environments, made France a crossfire zone of influences forcing workers to continually adapt or reinvent measures.[5]

As a result, the evolution of French, and indeed European, weights and measures—the units, standards, legislation, and administration—is intertwined with every aspect of European history and its commercial, industrial, and scientific dominance. Several eminent historians have struggled with this convoluted story, including the American historian Ronald Zupko and the Polish economist Witold Kula, who focuses on France in his *Measures and Men.*[6] Kula's book in particular depicts this story as a mirror of the character and vitality of premodern European life. Those who see no sense in European measures, according to Kula, cannot possibly understand Europe itself.

Farmers, for instance, had to cultivate enough land to sustain their families. Most European countries therefore developed a name for the amount of land that a farmer could plough in a day with a single ox or horse, or with a team of them, the measure sometimes called a *journal* (*jour* being French for "day"); in Lorraine, a *hommée* (man) was the amount of land that a man could work in a day. But the size of the unit depended on the crop; Catalonia had one journal for cornfields and another for vineyards. In Burgundy, the peasants measured "cornfields by the *journal*, vineyards by the *ouvrée*, and meadows by the *soiture*," all related to the labor put in them.[7] Varieties in climate and land quality also affected the size of these units.

Other kinds of labor developed their own measures. The "wash" was a capacity measure in one part of coastal Britain, referring to the amount of oysters washed at a setting; the "werkhop" in a different part of England referred to the grain harvest from a day's thrashing; and the "meal," in still another locale, referred to the volume of milk that could be gotten from one cow at a single milking. Local conditions produced variations. Wine was sold by the cask, and in areas where the wine tends to spoil quickly, such as Languedoc, the casks were smaller.[8]

Transporting goods spawned units. Goods that needed to be transported on the backs of animals came to be measured in bags, sacks, or packets—the size depending on the animal, the goods carried, and the distance. Goods hauled by other means could be measured in cartloads, wagonloads, and boatloads, or in the barrels or casks specially made to fit in such vehicles. Evolving needs and technology— new markets, better transportation—reshaped old units and created new ones.

Additionally, French farmers, merchants, and laborers worked in the midst of an everchanging political and institutional context much different from elsewhere in the world. Kula wrote, "Thus, we know of situations where, within a single village, one measure was used in the market, another in the payment of church tithes, and yet a third in rendering dues to the manor. Such arrangements were quite usual in the context of feudalism and similar social structures and, in principle, need entail no abuses or protests."[9] As these social structures gained or weakened in strength, the dominant measures could change accordingly; when states were politically strong and unified it tended to consolidate and simplify measures, when weak and fragmented it tended to multiply them. "[T]raditional measures," Kula wrote, "were firmly rooted in the realities of life and labor . . . the apparent chaos obeyed strict, long-established, 'organic' rules, which left no room for arbitrary conduct."[10]

In premodern times, "realities of life and labor" included oppression and exploitation, and efforts to combat them. A buyer might get more grain into a capacity measure by heaping or patting it, by pouring it from shoulder level rather than lower, or by pouring it in a busy mill, whose shaking tended to make the grain settle. Strategies to curb these abuses might include "striking" or leveling off the grain, insisting that it be poured from a dropped-arm position, or avoiding mills while their machines were operating. When abuses got out of hand, authorities might impose a variety of moral, religious, and legal sanctions. In Gdansk, Kula wrote, a person could have his fingers cut off for dishonest measures, while in thirteenth-century Latvia one could even be put to death for falsely measuring an ell, a length measure—but only if the amount of cheating was greater than two finger breadths.[11]

Measurement methods lasted because they were useful, customized for specific crops or goods and for local resources and social conditions. One area of land was not equal to another, due to the quality of soil and rainfall, making units derived from labor time and fertility more useful than neutral units with abstract lengths and weights. As conditions changed, so did units. Kula saw the history and culture of Europe reflected in the complex, dynamic ways it measured:

> It [the European system] reduced to common measures nature and culture the world around, and man's artifacts as well. It not only enabled man to measure fields, trees, and roads, but also imposed its proportions on the dimensions of the weaver's loom, and on bricks and church belfries, the dimensions for the bricks being part of the same system as the proportions of the church architecture. . . . [Such] measures, which had begun to evolve in prehistoric times and which had been endlessly

improved over thousands of years, when once they evolved into a coherent system, served man well in his work. They enabled him to satisfy his daily wants and to create immortal works of art: the noble proportions of the Romanesque, Gothic, and baroque cathedrals still astound us today.[12]

THE SUDDEN TRANSFORMATION

Then, in a staggeringly short period, the country spearheaded a revolution in measurement in the wake of the 1789 French Revolution. The roots of this revolution, however, can be traced back to a network of factors—social, political, scientific, and technological—that began to take effect at the outset of the seventeenth century. A new Europe was fomenting, quite different from the one Kula studied and appreciated. In it, not only were the old measures a burden, so was the entire measurement system. The eventual outcome would be a new and universal system, one that—ironically—only France could have provided.

Technologically, European industrial workshops were becoming increasingly stocked with and dependent on machines, from clocks to printing equipment, naval gear to cannons, spinning jennys to steam engines. The tools and equipment used to produce these machines required ever-better precision to build and maintain. Machines have parts, parts break, and replacements must have the same precision as the old. The proliferation of new and more complex machines made precision manufacture critical. Machines with interchangeable parts—clocks in 1710, muskets in 1778—brought this demand to a new level: parts had to fit not just one machine but any machine of that type. A new profession emerged of scientifically trained mechanics who specialized in precision engineering.

This demand for precision parts required precision measuring devices. The measuring network began to rule over the individual elements in it; the functionality of measuring became a property, not of this specific figurine or pitch regulator, but of the entire network of measuring elements. The growing extension of the metrological network, and the growing anonymity of its parts, went hand in hand with the loss of the social richness of individual measurement acts. Nothing like Niangoran-Bouah's play could occur in these workshops, where a measurement act was just one element in an extended social encounter between buyer and seller. Purchasing food, constructing buildings, and replacing parts were more and more automatic and anonymous. Crafts were being replaced by commodities. The famous factory discussed by English political economist Adam Smith in *The Wealth of Nations* (1776), in which division of workers' labor made it possible to produce hundreds or thousands of times more pins than before, was built with standardized processes. These metrological developments were therefore not just technological, but part and parcel of the new economic and political milieu in the emergence of capitalism.

Political changes, meanwhile, transformed the administration of weights and measures. In the Middle Ages, local manors and fiefdoms could get away with having their own measures and ignoring the central government's orders; central governments were relatively small and lacked an extensive bureaucracy. That began to change in France in the seventeenth century. The waning power of the feudal lords marked the disappearance of a key source of resistance to unification of weights and measures. Expansion of national and international markets was an incentive for central governments to insist on commonly agreed-upon measures and on controlling their supervision. "The equalization and standardization of weights and measures," writes the economist Stanislas Hoszowski, "stand in direct relation to the range of exchange relations (commerce) between given territories."[13]

Socially, the idea of a national identity was increasingly intertwined with uniformity of weights and measures. The replacement of a hierarchical sense of social authority with an intangible but growing sense of a shared "brotherhood of man" was allied with a sense that measures should also be shared and equitable, and men freed of the measurement exploitation common in the Middle Ages. Kula insists that this shared brotherhood was required before premodern, agrarian European nations could significantly alter their metrological systems: "Measuring reform cannot be achieved without a prior Declaration of the Rights of Man and Citizen, without first abolishing feudal rights, and without a well-developed market economy." It works the other way as well, for such glorious ideals would be impossible to sustain without metrological reform. "The increasing standardization of measures through time," says Kula, "is an excellent indicator of one of the most powerful, if, indeed, not the most powerful, historic processes—the process of the waxing unity of mankind."[14]

Scientific thought was also dramatically changing. Until late in the Middle Ages—following Aristotle—the universe was generally viewed as a cosmic ecosystem which included vastly different regions—the heavens and the earth, first of all—containing different kinds of things to which different kinds of generalizations applied, and to which different measures were appropriate. Local places called for local measures. The science was qualitative; rules were generalizations of how nature usually works as we humans usually experience it. All this changed in the early modern era. Nature came to be described not by rules but laws, produced not by generalizations but measurements. Galileo's first piece of scientific writing, composed when he was only twenty-two, was "The Little Balance" (1586), which described his improvements on a common measuring instrument and its use for defining the relative densities of substances. By the time of Newton's *Principia* (1687) a century later, the idea of local places had been superseded by a concept of "space" as single and uniform. The heavens and earth were not different places made of different kinds of material obeying different rules; they belonged to one space and obeyed one set of mathematical laws. This space—the world—is a stage on which only measurable pieces of matter appear, moved by measurable forces in measurable motions. Nature was understood not by considering the various overlapping roles that its objects play in a cosmic ecosystem, but by detaching them from their place and understanding them by their location in space-time on the world-stage. An abstract space requires an abstract measure. Not only equality of human beings but also abstraction of space is required. Distinctions between regions, products, and times cannot leave their imprint on a measure, in what Kula, with regret, calls "alienation of the commodity." This new world is measurable, calculable, and universal. Nothing is final, the world is open-ended, and anything can be measured and remeasured with infinitely greater precision. For the metric system's victory, Kula says, "two conditions had to be satisfied: the equality of men before the law, and the alienation of the commodity."[15]

The Soviet writer Ilya Ehrenburg wrote in his memoirs of dining after the First World War with the French writer Georges Duhamel, who declaimed enthusiastically about the metric system. "When Duhamel left, we broke into laughter," Ehrenburg wrote. "We liked his books, but his naïveté amused us: he was apparently convinced that he could measure our roads with the metric yardstick."[16] Places, each different, have to be estimated differently; these were giving way to Space—one, uniform, and measurable in only one way.

The impact of this abstract character of space in the new science of the seventeenth century is well captured by Harvard historian of science Steven Shapin. Referring to an episode when the French philosopher-scientist Blaise Pascal sent his brother-in-law up the Puy-de-Dôme peak with a barometer to see what would happen, Shapin comments: "In practice, the natural philosopher does not care what happened to *this* mercury in *this* piece of glass apparatus on *this* day and at *this* place, except as these outcomes support inferences to the relatively nonlocal and nonspecific. The local and the specific are not the point of these experiments."[17]

In 1690, the British philosopher John Locke spelled out the implications of this idea for measurement. A length, he wrote, is just an idea we have of a certain amount of space. Once we humans have that idea we can apply these lengths to measure bodies regardless of their kind or size by adding one specific segment to another. Measures have no necessary connection to human parts or purposes; they are abstract determinations by abstract ideas. Measures should not emerge from workshops and the tasks performed there, but should be fashioned by the human mind.

This notion reinforced the case for using decimal scales in measurement, rather than traditional fractional divisions. Fractional divisions are practical in the marketplace; it is easy to eyeball half or twice a certain magnitude, and half or twice that again. Like everything else about weights and measures, the decimal system was not natural but a human creation. The Chinese used a decimal system already in the twelfth century, but it was not even proposed in Europe until the end of the sixteenth. Because abstract calculation is easier with decimals, scientists were quick to embrace it. Numbers acquired a new importance in science. "[B]efore the age of Kepler, Galileo, and Harvey," writes the historian of science I. B. Cohen, "numbers were not used to express general laws of nature or to provide testable questions to test a scientific theory. This feature of the use of numbers in science set the new science of the Scientific Revolution apart from the traditional study of nature; in fact it defines the newness of the new science."[18]

Scientists discovered, however, their needs for precise measures and instruments outstripping what was available. Galileo struggled to measure time precisely enough for his studies of pendulums and balls rolling on inclined planes, William Harvey had difficulty measuring blood flow in his studies of the circulation of the blood, and Johannes Kepler recognized the value of precise astronomical measurements while working on a way of characterizing the heavens superior to that of Ptolemy. In this way they and others used measurements to describe phenomena in ways scientists never had before. Scientists were finding themselves hampered by poor and inconsistent measures and by the different measures used by colleagues in different lands; they were interested in improving measures, and inventing and refining measuring instruments and methods.

At first, scientists met in informal groups, such as the Accademia dei Lincei to which Galileo belonged. In France, a community of scientists whose early members included René Descartes (1595–1650) grew more assertive, and in 1666 Louis XIV sanctioned the French Academy of Sciences. With royal backing, they were able to embark on ambitious projects such as studies of the shape of the earth. In England, another group of scientists, initially collecting around followers of Francis Bacon (1561–1626), grew into the "Royal Society of London for the Promotion of Natural Knowledge," which was officially incorporated in 1662.

The French Academy and the British Royal Society soon began to collaborate. Members of both institutions endeavored to find unchanging phenomena that could be used to evaluate the accuracy of standards and to re-create standards if they were damaged, lost, or destroyed. There were two principal candidates. One was the "seconds pendulum," or pendulum whose bob took a second to swing once in either direction. Galileo had discovered that the time of oscillation of a pendulum depends only on its length, meaning that the length of carefully built seconds pendulums undisturbed by other influences should always be the same anywhere on Earth. As it happened, a seconds pendulum is about a yard long, a convenient length for a measurement standard.

The other candidate was the earth's meridian, or great circle passing through both poles. The circle around the equator was occasionally proposed, but would be more difficult to measure and only passed through certain countries—while every nation has meridians passing through it. The earth's meridian would be difficult to measure accurately, but was sure to remain constant.

Scientists assumed that available technology would soon measure the seconds pendulum and the meridian accurately enough to define a unit, so that if a standard would be damaged or destroyed, it could be re-created with the same or greater precision. This assumption would prove false. Yet in the seventeenth and eighteenth centuries, scientists' sheer confidence in it stimulated the drive for a universal metrological system.

In 1670, Gabriel Mouton (1618–1694), a founding member of the French Academy, found that seconds pendulums varied with latitude due to deviations in the earth's shape from a perfect sphere. He used these variations to recalculate the length of the meridian and proposed its subdivisions be used as fundamental length measures.[19] Mouton called the minute (the sixtieth part) of an arc a *mille* or milliare, with other units being decimal fractions of it: thus one-tenth was a stadium, one-hundredth a funiculus; one-thousandth a virga, one ten-thousandth a virgula, and so forth.[20] The virga and virgule were roughly the size of the toise and pied of the existing French measures.

Mouton's colleague Jean Picard (1620–1682), another Academy cofounder, organized an expedition to measure the arc of the meridian that passes through Paris. As a first step, in 1668, he helped to fix an iron bar fastened to the outer wall of the Châtelet palace, which had long been used as a measure of the toise and was in disrepair. His new estimate of the meridian—57,060 toises, measured according to the new

Châtelet standard—was vastly more precise than previous measurements and ushered in a new era of geodesy, or earth measurements.[21] Picard also proposed a universal standard.[22] The deterioration of the Châtelet toise indicated the need for something permanent, and the seconds pendulum offered an opportunity. Picard measured the length of a seconds pendulum and found it to be 36 inches, 8½ lines per the Châtelet toise. The pendulum varied slightly with the seasons, due to the effects of changing temperature and humidity, so these factors had to be taken into account. Once they were, the seconds pendulum amounted to an "original [measure] taken from nature itself, which ought therefore to be invariable and universal," and for standards we "will need no other original but the Heavens." Picard called the length of a seconds pendulum the "Astronomical Radius," a third of it the "universal Foot," its double the "universal Toise," its quadruple the "universal Perch," and 1000 such perches the "universal Mile."

The idea that either the meridian or the seconds pendulum could be used as a natural standard attracted more attention. In 1720, Giacomo Cassini (1677–1756), the second of four generations of astronomers and often known as Cassini II (his father, Cassini I, had come from Italy to become the first director of the Paris Observatory) further developed Picard's meridian measurements, carrying the arc measured north to Dunkirk and south into Spain. Cassini II proposed a "geometrical foot" that would be 1/6000 the minute of arc; six such feet would be a toise.

Cassini's measurements suggested the earth's shape was a prolate spheroid—egg-shaped, or thinner at the equator than at the poles. This contradicted Newton's conclusion that it was an oblate spheroid, flatter at the poles from centrifugal force than at the equator. In 1735, the French Academy sought to resolve the dispute by mounting a new expedition to measure the length of a meridian in Peru, close to the equator, and in Lapland, close to the poles. In preparation yet another, more

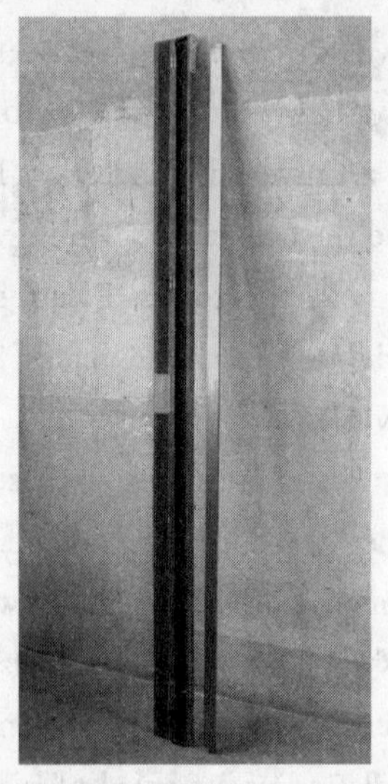

THE PERU TOISE, A FRENCH LENGTH STANDARD OF THE 1730S.

careful toise standard was constructed. This effort, which determined Newton to be right after all, created a French length standard of a new precision, known as the Peru Toise. It was divided into 6 feet, each foot into 12 "thumbs," and each thumb into 12 lines. In 1766 eighty copies were made and sent to various parts of France, including the Châtelet.

Other eighteenth-century proposals to tie a length standard to either a seconds pendulum or some fraction of the earth's meridian included "New Attempt at an Invariable Measure Able to Serve as a Common Measure for All Nations" (1747), by French Academy member Charles Marie de Condamine, who argued for the seconds pendulum. In 1773, Jean Antoine Condorcet (1743–1794), the new assistant secretary of the French Academy, began working with Anne-Robert-Jacque Turgot (1727–1781), the controller general of finance under Louis XVI (1774–77), on a proposal for a uniform system of weights and measures. It called for a length standard consisting of "the length of the simple pendulum which vibrates seconds in a given latitude" ; the standard weight would be "determined in like manner by philosophical methods," such as by creating a cubic vessel of a standard shape according to the length standard and weighing the amount of pure water that filled it.[23] The plan was shelved in 1775 when political intrigue cost Turgot his position.

In England, Royal Society members also explored tying measurement standards to the seconds pendulum and the meridian, and eventually collaborated with their French colleagues. Christopher Wren (1632–1723) in England and the Dutch physicist and mathematician Christiaan Huygens (1629–1695), a member of both the French Academy and British Royal Society, proposed the seconds pendulum as a standard. In 1742, in a milestone of scientific collaboration, the Royal Society made two copies of a linear measure on which it marked off its standards and sent both to the Academy for the French to mark off their standards; the Academy kept one and returned the other.

In 1758, the Royal Society worked with the members of a royal commission, known as the Carysfort Commission after its chairman, which carried out an extensive review of existing standards and created new ones. "[F]or no less than 415 years," its report said, "that is, from the Great Charter down to the 16th of Charles I. the Statute Book abounds with Acts of Parliament, enacting, declaring, repeating, that there should be one uniform Weight and Measure throughout the realm; and yet every statute complains that the preceding statutes were ineffectual, and the laws were disobeyed."[24] New British linear and weight standards were created in the wake of the committee's report. But legislation proposed in accord with its recommendations went nowhere.

Such proposals, even influential ones, were doomed without political muscle. This emerged in France in the second half of the eighteenth century, in the decades leading up to the French Revolution, thanks to disruptions of commerce, popular unrest, and scientific pressure. The controller general of finances considered metrological reforms in 1754, and 10 years later in 1765. In each case the king demurred, in view of the cost, the struggle it would take to alter custom, and the unpleasant confusion it would no doubt create. Popular pressure mounted, but the successive kings kept dragging their heels. In 1778, Finance Minister Jacques Necker (1732–1804) wrote a report for Louis XVI pointing out the advantages and disadvantages of radical metrological reform, but the report's conclusion found with the disadvantages.

Yet pressures for reform mounted. Matters came to a head in 1789, when Louis XVI asked the three estates or divisions of French society—the nobility, clergy, and commoners—to express their grievances directly to him. The result was a collection of grievances known as the *Cahiers de doléances*, "the most extensive investigation of public opinion in Europe until our own century, and a detailed expression of French social life on the eve of the Revolution."[25] Many cahiers complain bitterly about the existing system of weights and measures, which had been used ruthlessly as a tool of abuse by landlords over peasants; these cahiers demand "one weight, and one measure."

FRENCH REVOLUTION

After the storming of the Bastille, a fortress in Paris, in July 1789, the diversity of weights and measures and their accompanying abuses—already a huge source of discontent—became a political symbol. French Academy members began discussing among themselves how best to have their institution weigh in, so to speak. In August, Academy member Jean-Baptiste Le Roy (1720–1800) proposed that they petition the National Assembly to institute a uniform standard for weights and measures.[26] The Academy members debated which National Assembly member would be best to invite as a spokesman for the cause. They settled on Charles Maurice de Talleyrand-Périgord (1754–1838).

Talleyrand was a wise choice, a canny politician of aristocratic background who knew how to ride currents and cut losses. Ordained

CHARLES MAURICE DE TALLEYRAND-PÉRIGORD.

a priest in 1779, he became the Church's representative to King Louis XVI the following year. When revolutionary fever swept France, he joined it enthusiastically, helped write the Declaration of the Rights of Man, turned against the very church to which he owed his high political profile, and was excommunicated. Meanwhile, he recognized from the cahiers the urgency—and powerful symbolism—of reforming weights and measures.

In 1790, after consulting with Academy members, Talleyrand presented a proposal to the National Assembly. "The great variety in our weights and measures occasions a confusion in our ideas, and necessarily an obstruction to commerce," he began. Abuses are rife, and it is the National Assembly's duty to intervene. He reviewed French failures to unify weights and measures—but today, he said, we are in a "more enlightened age" and can meet the challenge. The easiest and simplest path would be to adopt the existing Paris pound and toise—but better yet to be more ambitious, for scientists have shown us how to base measures on an "invariable model found in nature," so that when standards are lost or damaged they can be replaced. He proposed to define the aune as the length of the seconds pendulum; the toise as its double; and to divide the toise into feet, inches, and lines. The eminent French scientist Antoine Lavoisier (1743–1794), he said, will study how to derive a weight measure by taking the weight of a cube of water whose side is one-twelfth the length of this pendulum. England, he continued, will surely join France in this cause of reform, "in which our commercial connections give us a common interest; and which may hereafter be beneficial to the whole world," and equal numbers of the members of the British Royal Society and French Academy of Sciences should arrange to meet at some suitable place to deduce an invariable standard. In this way, France would pioneer a "political union brought about by the mediation of the sciences." Changing measures will bring "some disorder," but once everything is explained, and conversion tables distributed, a mere 6 months will be needed to mandate the system.[27]

The National Assembly approved Talleyrand's proposal, and on August 22, 1790, so did Louis XVI. Popular passions were continuing to run high over the need for metrological reform, and even provoking riots; traditional measures had been a tool of abuse by which the old feudal system oppressed peasants. Time for a modern and scientific system! This gave Talleyrand's proposal all the momentum it needed.

The king sent Talleyrand's proposal to the Academy, where one committee recommended that it be based on a decimal system and another investigated natural standards. Lavoisier assisted both committees. On March 19, 1791, the second committee issued its report, "On the Choice of a Unification of Measures," which outlined three possibilities for a natural standard: the length of a seconds pendulum, a quarter of the earth's equator, and a quadrant of the meridian running through

Paris. The committee held for the third: the basic length unit would be a ten-millionth part of the Paris meridian.

The choice required mounting a new expedition to measure the meridian arc, which would substantially boost the Academy's budget, prolong its involvement in the development of the system, and provide an opportunity for additional scientific measurements and tests of scientific equipment. Also, given that the meridian had already been measured half a century before, the result would not be a surprise and a crude provisional standard could be processed immediately.

The Academy wanted to create a standard that all nations might be persuaded to adopt: "The Academy has done its best to exclude all arbitrary considerations—indeed, all that might have aroused the suspicion of its having advanced the particular interests of France." They soon adopted the name "meter" for the basic length unit, stemming from the Greek word *metron*, meaning "measure."[28] The Greek name also helped make the product sound more universal than French. The Academy then set out to create a decimal system of length measures based on divisions and multiples of the meter. Capacity units would be produced by cubing these length measures; weight units by filling these capacity units with distilled water. Length, capacity, and mass units would all be linked together, the entire system derived from one universal and unchanging standard.

Talleyrand took the plan back to the Assembly, which approved it on March 30, 1791. The Committee of Public Enlightenment, swelling with a mixture of patriotic fervor and rationalistic universalism, proclaimed: "Surely this shows . . . that in this field, as in many others, the French Republic is superior to all other nations."[29]

Academy members formed several committees to carry out the project. One, entrusted to Pierre Méchain and Jean Baptiste Delambre, was to conduct the meridian measurement. Another, to measure a seconds pendulum at 45 degrees at sea level, was given to Jean-Charles de Borda, Méchain, and Jean Dominique Cassini (Cassini IV). A third committee, to which Lavoisier belonged, was to discover the weight of a certain mass of distilled water at the freezing point, which would help determine the standard weight. A fourth committee was assigned to compile tables of equivalents between the new system and the old.

It was a difficult task, carried out while the idealism of the French Revolution was disintegrating into a series of brutal and bloody dictatorships. Power was held by increasingly radical and often ruthless legislative bodies—the National Assembly (1789), the National Constituent Assembly (1789–1791), the Legislative Assembly (1791–1792), the National Convention (1792–1795; its Committee on Public Safety exercised power during the Reign of Terror from June 1793–July 1794), and the Directory (1795–1799). The project would take over 7 years to complete.

On June 19, 1791, a dozen Academy members, including Cassini, met with King Louis XVI to discuss the project. The king, mystified, asked Cassini why he wanted to measure the Paris meridian again, given that his ancestors had already measured it. Cassini patiently explained that modern instruments made it possible to improve the measurement. The king was surely distracted, for the next day he attempted to flee the country with Marie Antoinette and their son, the dauphin, but was captured and imprisoned. His endorsement of the Academy's measures project was his final act as a free official.

Méchain and Delambre set out on the expedition later that year, taking two platinum standard rods that were double the length of the Peru toise. Their mission turned out to be difficult and dangerous. The Revolution was still in progress, unpopular in areas, and Méchain—an official representative of a revolutionary government—was detained as a spy in Spain. Their epic story is recounted in the book *The Measure of All Things: The Seven-Year Odyssey and Hidden Error That Transformed the World*—the "error" mentioned in the title, a little breathlessly and hyperbolically, refers to mistakes made in this work by Méchain and then covered up by himself and Delambre.[30]

Meanwhile, in France, conditions were horrifying. The revolutionaries, exhibiting decimal fervor, ordered clock time decimalized: 10-hour days, 100-minute hours, 100-second minutes. This action rendered every existing watch and clock useless, appalled otherwise sympathetic countries, and was soon abandoned. A calendar mandated that year renamed the months and restarted Year One in 1792. This reform managed to last 10 years. By this time, the guillotine had become the revolutionaries' tool of choice for executions, and King Louis XVI was beheaded in January 1793. The National Convention's Committee on Public Safety, established that April, would serve as the executive branch of the government during the ensuing, year-long Reign of Terror.

One constant throughout this tumultuous and tortured period was widespread enthusiasm for the metric system. It seemed a clear and achievable goal, and an indispensable element of the rational, egalitarian, and universal society which the revolutionaries aspired to establish. On August 1, 1793, the National Convention approved the French Academy's plan, putting into effect the new system based on a ten-millionth part of the quadrant as it had been determined in 1740—but mandated use of the new system by July 1, 1794. This hopelessly optimistic deadline was virtually impossible to meet. The Academy's project was not yet complete, many names of units were not yet established—though "meter" was—and standards had yet to be developed.

Worse news followed. Academies and literary societies of all kinds fell under suspicion and were abolished by a decree of August 8, 1793. Lavoisier lobbied for the Academy's continuance and Antoine Fourcroy (1755–1809), another Academy member, managed to have a temporary commission of weights and measures created on September 11, 1793.

But 3 months later, the Committee on Public Safety purged six members of the commission for being insufficiently revolutionary, including Borda, Coulomb, Delambre, and Lavoisier. Condorcet fell under suspicion because of a political pamphlet he had published, and his arrest was ordered. He fled and hid underground for months. He was finally arrested on March 27, 1794, taken to a prison in a suburb of Paris newly renamed the Bourg-Egalité, or "Equalityville," and found dead in his cell the next morning, with some people at the time suspecting suicide. Lavoisier, one of the greatest scientists of all time, was imprisoned in November 1793 and guillotined on May 8, 1794.

Nevertheless, the Committee on Public Safety was determined to establish the new system of weights and measures and provide them to the world, and on December 11, 1793, it ordered a doctor and scientist named Joseph Dombey to take provisional standards to the United States. The metrological reform had several factors going for it. One was the zeal of the revolutionaries, who had the political will and power. For them, reform was part and parcel of the overthrow of feudalism and the ancient regime, and essential to the creation of liberty and equality and to the expunging of serfdom. Citizens were ordered to use the new system to prove their civic loyalty. Another was that, unlike the clock and calendar reforms, metrological reform was urgently needed.

The temporary commission dutifully created an instruction manual for the new system, published in 1794. But on April 7, 1795, the National Convention withdrew the previous law of August 1, 1793, replaced most of the names, and announced that it would make the system mandatory—despite the fact that there were as yet no standards. The basic names—meter, liter (a cubic decimeter), and gram (defined as the weight of a cubic centimeter of pure water at the temperature of melting ice)—would be used to build a comprehensive nomenclature, with prefixes to denominate ten, hundred, and thousand, and tenth, hundredth, and thousandth.

PROVISIONAL METER STANDARD AT THE PLACE VENDÔME, ONE OF MANY PLACED FOR PUBLIC USE THROUGHOUT PARIS BETWEEN 1796 AND 1797.

This would make it easy to move from one scale to another; it would be simply a question of moving a decimal point. But the Convention did not set a date for when the old system would be abolished. This law was the closest to the effective birth of the metric system itself, if not yet of its standards. Copies of the meter were placed all over Paris to familiarize the public with the new standard; of these, only two are left.

Meanwhile, the measurement of the meridian arc had been suspended because of Delambre's purge from the commission and Méchain's detention in Spain. On April 17, 1795, the government appointed a committee to restart the meridian arc measurement, and in October created a new institute to replace the Academy. That October the Convention was dissolved and taken over by yet another legislative body, the Directory.

Fortunately, Delambre was rehabilitated and Méchain released, and the two returned to their measurement of the meridian arc. By January 1798, as the various committees were finishing their projects, several institute members suggested inviting scientists from other countries to participate in the completion of this system, which was after all intended for the use of the entire world. Talleyrand agreed and sent invitations to several neighboring countries known to be sympathetic to the ongoing Revolutionary government. Foreign delegates began arriving in September—eleven in all from Spain, Denmark, and various European republics—who joined ten French scientists to study the problem of creating new standards. Méchain returned in November 1798, and the committee—whose foreign participants outnumbered the hosts—met on November 28. It was a scientific gathering of a new kind, writes historian Maurice Crosland, "a transition toward the modern idea of an international scientific congress."[31]

The conference participants wrote up a report on April 30, 1799. They had determined the length of the meridian quadrant to be 5,130,740 toises, and they based the length of the meter on this figure. A measurement of distilled water was used to establish the weight of the kilogram standard. The new standards were fashioned out of platinum, with the meter standard, called an étalon, having a rectangular cross-section, 25 millimeters wide and 4 millimeters deep. On June 22, 1799 the meter étalon and the kilogram were officially presented to the legislature. An unidentified spokesman—probably Pierre-Simon Laplace (1749–1827)—gave a moving tribute to the Academy's effort on the occasion. Our work, the spokesman said, so useful to the entire world and such a reflection of the glory of France, is now complete. Until now, every country's measures have been arbitrary. Now, thanks to the Academy's efforts, the world has a measurement system based on Nature herself, as unalterable as the globe. Everyone on earth will be able to understand and experience an affinity with the new system. Fathers can say, with pleasure, that "the field which feeds my children is part of that Globe; it is to that extent that I am co-owner of the World." The spokesman went to great pains to emphasize the international character of the system, pointing out that the Academy had invited many foreign scientists to participate in its creation—and when he called out the names of the conference participants, he did so in alphabetical order, giving no priority to the French scientists. Even if some catastrophe should swallow the earth, or some stroke of lightning destroy the standards we have created, our work is not in vain, for we have also tied the standards to the length of a seconds pendulum in Paris, meaning that we could reconstruct the standards which we now present—a "meter *from* nature *for* measuring nature, and a true kilogram derived from that." We shall protect two standards, the spokesman concluded, "with a religious vigilance."[32] Ever since, these two standards have been known as the Meter—and Kilogram—of the Archives.

It was indeed a landmark moment for metrology, for science, and for civilization. In a report written two decades later, in 1821, John Quincy Adams would write of the event:

> The spectacle is at once so rare and so sublime, in which the genius, the science, the skill, and the power of great confederated nations are seen joining hand in hand in the true spirit of fraternal equality, arriving in concert at one destined stage of improvement in the condition of human kind; that, not to pause for a moment, were it even from occupations not essentially connected with it, to enjoy the contemplation of a scene so honorable to the character and capacities of our species, would argue a want of sensibility to appreciate its worth. This scene formed an epocha in the history of man. It was an example and an admonition to the legislators of every nation, and of all the after-times.[33]

• • •

Le Brigand showed me a few other artifacts of the Revolution stored in the safe. These included thermometers used to standardize the temperature scale and copies of the meter and kilogram prototypes to be used for comparisons and for producing other copies; these copies, unlike the meter étalon itself, were marked off in centimeters, decimeters, and millimeters. The meter copy had dark smudges in several places, what the touch of an oily finger on platinum looks like 200 years on.

But my attention kept returning to the featureless originals, bearing no references to rulers, dates, or to the cultural or natural environment. I thought of the West African gold weights, the Chinese flutes, and Pile de Charlemagne I had seen in a case adorned with fleur-de-lis. These had been carefully and individually crafted not only to be used but also to be visibly appreciated. These artistically designed measures had been replaced by the faceless prototypes that Le Brigand was now putting away, the heavy metal doors screeching again as they were shut. It was a dramatic change in measurement and also in the role of measurement in the world.

In the words of the often-repeated public slogan: "A tous les temps, a tous les peoples" (for all times, for all peoples). But it would take some time for the other peoples to agree.

CHAPTER FIVE

HALTING STEPS TOWARD UNIVERSALITY

On January 17, 1794, a French doctor and botanist named Joseph Dombey stepped aboard the *Soon*, a brig departing from Le Havre for Philadelphia. Dombey bore a letter of introduction from the Committee on Public Safety, the executive body that ruled France during the Reign of Terror. Dombey was carrying to the U.S. Congress a provisional copper length standard—newly named the *mètre*—and a copper weight measure, not yet officially named the *kilogramme*, intended to help the United States reform its system of weights and measures.

The French Revolutionaries had chosen their emissary well. Dombey had an engaging personality and a wealth of scientific learning that would surely impress the Americans. "He had integrity, courage, and a sense of adventure," writes historian Andro Linklater in *Measuring America*. "He was the ideal choice in every way but one—his luck was phenomenally bad."[1] Indeed, his story was so calamitous that, had it taken place any earlier in history, surely it would have provided the material for an operatic tragedy, for a farce, or both.

As a young man, Dombey (1742–1794) was an avid student of medicine and natural history, and became a physician. In 1776, at age thirty-four, he was assigned to a Spanish botanical expedition to South America, during which he built up France's collection of botanical specimens of South American plants, earning him a seat in the Academy. His experiences on this excursion were challenging—he got dysentery and was forced to delay publishing his findings until after his Spanish colleagues. Disgusted with the politics of botany, he retired to Lyons to practice medicine in a military hospital.

Not a good choice. During the Revolution, Lyons was an enclave of resistance to the Reign of Terror, and its inhabitants were attacked and humiliated by the revolutionaries. Dombey watched his patients dragged from the hospital and guillotined. Worried about his sanity, some well-connected friends arranged another expedition for him—to the United States, to bring samples of the new, rational system of weights and measures system to their ally and to collect botanical specimens.

Dombey never made it to American shores. In March, as the boat neared Philadelphia, a fierce storm damaged the brig and drove it south to the Antilles, where it landed at Point-à-Pitre in Guadeloupe. This French colony was as politically divided as France itself. Its governor was royalist, but Point-à-Pitre was full of revolutionary sympathizers. Dombey could not avoid becoming a political pawn. The presence of an emissary of the revered Committee on Public Safety from the home country inflamed the fervor of the locals against the governor, who had Dombey arrested and imprisoned. A mob gathered to demand the release of the man who was, in truth, an official representative of the French government in a French land.

Dombey's release incited the mob to take revenge against his captors. Standing on the bank of a channel, Dombey tried to stop the violence, but was pushed off the bank into the water. He was unconscious when fished out, and caught a raging fever.

The governor took Dombey into custody, interrogated him, realized he was no agitator, and put him back aboard the *Soon*. Right after the ship left the harbor, it was attacked by British privateers who seized its cargo and took the crew hostage. Though he disguised himself as a Spanish sailor, Dombey was recognized and imprisoned for ransom at the British colony of Montserrat, where at the end of March—still ailing—he perished.

Back in France, during the darkest days of the French Revolution, no one at the Committee on Public Safety was concerned by the absence of news from Dombey; they learned of his fate only months later, in October.

The *Soon*'s cargo was auctioned. Dombey's meter and kilogram were purchased by someone who sent them to a French official in Philadelphia. This official put them into the hands of someone who, not realizing their significance, never conveyed them to the U.S. Congress. Had Dombey's mission succeeded, it might have provided momentum to advance the metric system in the United States. "The sight of those two copper objects," Linklater writes, "so easily copied and sent out to every state in the Union, together with the weighty scientific arguments supporting them, might well have clarified the minds of senators and representatives alike. The vibrant, determined personality of Dombey could have created an immediate empathy. And today the United States might not be the last [major] country in the world to resist the metric system."[2]

Dombey's mission took place at a tumultuous time in the history of weights and measures. In China and West Africa, as in other parts of the world, Great Britain's aggression and outright colonization would soon start dislodging regional measurement systems as the country began to impose its own. In Great Britain itself, as well as France and the United States, radical reforms were either in the works or under serious discussion in a novel way.

Certainly, governments had revised weights and measures before. They had imposed systems of measurement on others. What was happening in the 1790s was new. Three major countries were seriously considering imposing, *on themselves*, a radically new system of weights and measures. The most dramatic features of this system were its decimal scale; its goal of tying the system to a natural standard; and the fact that the system would be supervised by scientists, not government administrators. John Playfair (1748–1819), a Scottish mathematician and geologist, declared: "The system adopted by the French, if not absolutely the best, is so very near it, that the difference is of no account."[3] Lavoisier said, "Never has anything grander and simpler and more coherent in all its parts come from the hands of men."[4]

Yet the efforts to sweep away entrenched systems faced numerous challenges. Success required pressing need, passionate leadership, and the right political climate. The effort had to be a crusade. In the 1790s, the metric system became one, for at least several key individuals in these three countries. The new system was advanced by its advocates as the most progressive, impartial way to measure nature in the Newtonian world of abstract space. Yet there was more than detached, scientific appreciation for it, there was zeal. Metric advocates saw the metric system as a necessary part of a truly enlightened world; opposition surely stemmed from dangerous vestiges of irrationality and superstition.

The metric system was now France's choice: would it be the world's? The world's communities had a wide range of measurement systems that were astounding in their originality and ingenuity, each tightly woven into the local culture. How did the French system achieve the universality that its planners envisioned?

GREAT BRITAIN

Great Britain was experiencing little revolutionary fervor around the end of the eighteenth century, and metrological reform made the least headway there. Parliament member John Riggs Miller (1744–1798) was a chief advocate of weights and measures reform. In July 1789, and again in February 1790, he made several impassioned speeches in the House of Commons urging the creation of a standard from something "invariable and immutable" in nature, which would be "at all times, and in all places, equal and the same." It is a government duty to make the means by which people "buy and sell, pay, contact, barter, eat, and live" so simple and self-evident "that the meanest intellect shall be at par with the most dexterous." Travel 10 miles, he said, and you find the sizes of acres, bushels, pounds, and gallons different. Who benefits? Only "knaves and cheats."[5]

Miller received a letter from the French minister Talleyrand, who had learned of Miller's ideas. "Too long have Great Britain and France been at variance with each other, for empty honour or for guilty interests," Talleyrand wrote. "It is time that two free nations should unite their exertions for the promotion of a discovery that must be useful to mankind."[6]

Emboldened, Miller delivered a third speech in Parliament on metrological reform, "its moral, its commercial, and its philosophical [scientific] influence upon mankind." He addressed the beneficial impact on trade as well as on social life. Regarding moral influence, complex and confusing weights and measures harm the honest and help the dishonest. In commerce, such measures disrupt trust, lead merchants to be overly cautious, and exposes them to lawsuits. Finally, scientific research is damaged by complex and inaccurate weights and measures. While small errors in commerce are not deadly and are even expected, small errors in science can be fatal.

Miller proposed four possible ways of deriving standards from nature. Two—the seconds pendulum and the fraction of a meridian—were well known. The other two were lesser known and ingenious: the distance bodies fell in a second, and a drop of water or wine at a particular temperature. Assuming drops to be of equal size, such drops could fix standard weights and volume—and even length if a cubic vessel were made to hold a certain amount, killing three birds with one stone. But this was a big assumption and probably impractical; therefore, Miller proposed the seconds pendulum. "By having an universal standard of weights and measures," Miller continued, "the traveler in all nations would be *at home* in respect to distances; the merchants of all nations would meet each other on known grounds in commerce, and the philosophers of all nations would do the like in science." Weights and measures should be established on a "permanent, unalterable foundation, whence invariable standards might be obtained, to which all nations might refer, and with which they might compare their respective measures, and reduce them to one invariable, universal denomination, for the mutual convenience and benefit of all mankind; by deriving them from such principles as might enable all future generations to obtain similar measures of length, capacity, and weight."[7]

The British, wary of French revolutionary fervor, were also wary of Talleyrand's attempts to enlist them in efforts to develop the new system. Moreover, British reformers took note of difficulties that the French were having in getting their citizenry to adopt their "rational" and "universal" system. Miller's proposals failed to rouse the populace, and he was defeated in the next election. Among the remaining promoters of metrological reform, George Skene Keith (1752–1823) advocated the seconds pendulum as a natural standard in his 1791 monograph, "Synopsis of a System of Equalization of Weights and Measures of Great Britain," and again took up the idea in 1817 in "Different Methods of Establishing an Uniformity of Weights and Measures." Francis Eliot (1756–1818), another reformer, authored the 1814 "Letters on the Political and Financial Situation of the Country," but his proposed reform died, though it did have the merit of introducing the name "imperial," soon to be adopted to designate the British system.

British reformers had nothing like the impact of their French counterparts. For one thing, they tended to act solo rather than cultivate government officials; they had no Talleyrand, nor revolution to give their proposals emotional and symbolic value. Finally, the British trade and economy were nowhere near as broken as France's. As Zupko writes:

> England was the ranking industrial nation in the world, and its manufacturing, commercial, and financial leaders argued that too abrupt or radical a change would impair present and future growth (a claim made repeatedly down to the present day). If machine parts had to be changed, or if the dimensions of most of England's exports had to be altered, untold confusion would result in the national economy and the entire nation would be thrown into a recession or a depression. (Similar arguments have been presented by anti-metric spokesmen in the United States throughout the present century.) It was far wiser, they claimed, to proceed cautiously with change and only modify those aspects of the existing metrology that were detrimental to further economic growth. London was convinced by these arguments (notwithstanding the political clout of such powerful lobbies) and by those coming from other sectors of the populace and eventually compromised with the legislation passed in 1824. Moderation rather than revolution was its theme.[8]

Moderation and delay. In 1814, a House of Commons committee recommended—finally—that the yard standard constructed for the 1758 Carysfort Commission be officially adopted. A second committee examined using a seconds pendulum as a natural standard, and the work was completed in 1818. The outcome was the Imperial Weights and Measures Act of 1824, effectively establishing the British answer to the metric system of a quarter century before.

The act created the imperial system of units, based on units inherited from the Romans, and used throughout the British Empire. The imperial system relates the length measures inches, feet, yards, and miles (63,360 inches in a mile), and the weight measures grain, ounces, pounds, and tons (14,000,000 grains in a "short" or ordinary ton). It was as revolutionary as the British got, legalizing work on standard weights and measures that had been carried out over a century before. Still, it tied a standard—the yard—to a natural phenomenon, the seconds pendulum, so that in case the existing standard were destroyed, it could be re-created. The pound was defined by a cubic inch of distilled water, weighted in air, at a temperature of 62°F and barometric pressure 30 inches. Both standards were stored in the House of Parliament. Finally, the most industrially advanced nation on Earth had weights and measures standards.

Then, on October 16, 1834, workmen burning old tally sticks in the basement of the House of Lords set fire to the building, in a conflagration that destroyed both houses of Parliament—and the recently enthroned imperial standards. This was exactly the kind of disaster that had made the idea of a natural standard such an appealing prospect; if a metrological system were pinned to such a standard, as the British scientists thought they had carefully done, the standards could be recovered exactly as they had been before with no change in the sizes of units.

A scientific commission, headed by George Airy, was assigned to find the best way of restoring the standards. Airy (1801–1892)—singleminded, determined, and orderly—saved everything, from correspondence to checkbooks. "He seems not to have destroyed a document of any kind whatever," wrote a biographer. He was "an organizer rather than a scientist . . . but he made great science possible"[9]—a remark that might apply to metrology itself. The biggest controversy in Airy's life involved a missed discovery of the planet Neptune. He had provided various scientists with information, and then happened not to be home when the chance came to make the discovery. He was "the prototype of the modern government scientist." Under Airy's careful guidance, the commission discovered far more uncertainties in the seconds pendulum than expected; the gravitational action, far from being universal, was sensitive to numerous slight variations and disturbances. It was next to impossible to ascertain the true length of a pendulum with anything near the precision of an artifact standard. After long and thorough deliberation, Airy's committee reported, on December 21, 1841, that the seconds pendulum method was not accurate enough to re-create the length standard. The commission set about to create new standards by using existing copies of the old ones instead.

The seconds pendulum had fired the imagination of those who believed in natural standards and universal systems for a century and a half. What should have been a triumph for metrology—the chance to prove that standards could be re-created if destroyed—instead turned into a major embarrassment.

UNITED STATES

Reformers in the United States, meanwhile, still had some momentum after its revolution of 1776–1781. A decade later, the country was still deciding its fundamental governmental and commercial structures, which had been largely inherited from the British. Leaders of the newly independent country were more than willing to change such institutions if better ones could be found. These institutions included the country's system of weights and measures.

Prior to the American Revolution, each colony determined its own weights and measures with little uniformity among them, much like the feudal regions of Europe. This created difficulties in interstate trade, among other issues. The Articles of Confederation of 1777, ratified in 1781, gave Congress the right of "fixing the standards of weights and measures throughout the United States" (Article IX), but no action was taken. Confusion springing from the diversity of weights and measures echoed what was going on in England and France. The new country employed the inherited British system, whose use was encouraged by the Land Ordinance of 1785 mandating land surveys. This despite the fact that the Ordinance's author, Thomas Jefferson, had been toying with the idea of a decimal system of weights and measures tied to a seconds pendulum. Had the Land Ordinance been delayed for a dozen years or so, the United States would have had an excellent opportunity to start from scratch in redefining its weights and measures.

In 1785 James Madison, then a member of the Virginia House of Delegates, wrote to fellow Virginian and Continental Congress member James Monroe, complaining of the coinage system and calling for members of Congress to intervene. Madison added that Congress should also address the separate but related issue of weights and measures:

> [W]ould it not be highly expedient, as well as honorable to the federal administration, to pursue the hint which has been suggested by ingenious and philosophical men, to wit: that the standard of measure should be first fixed by the length of a pendulum vibrating seconds at the Equator or any given latitude; and that the standard of weights should be a cubical piece of gold, or other homogeneous body, of dimensions fixed by the standard of measure?

Doing so, Madison continued, would not only be eminently practical and establish a uniform system throughout the United States but might also lead to a shared system throughout the entire world. "Next to the inconvenience of speaking different languages, is that of using different and arbitrary weights and measures."[10]

The Continental Congress did indeed shortly address the country's coinage woes; the dollar was officially adopted as the basic unit of coinage in 1785, followed by a complete decimal system a year later. Weights and measures remained unaddressed.

The founding fathers were still refining their thinking, and the desire to revise the Articles of Confederation in 1787 led to a thorough overhaul. The resulting Constitution, adopted in 1787, also gave Congress the power to establish "the Standard of Weights and Measures" (Art. I, Sec. 8). In 1789, the first U.S. Congress met and the country elected George Washington its first president. In Washington's first State of the Union address, on January 8, 1790, he stated that "Uniformity in the currency, weights, and measures of the United States is an object of great importance, and will, I am persuaded, be duly attended to." Following up, on January 15, the House of Representatives, meeting in New York, ordered Jefferson, now secretary of state, to prepare "a proper plan or plans for establishing uniformity in the currency, weights, and measures, of the United States."

Jefferson (1743–1826) was the right person to take this on. He was passionately interested in science, both theoretically and practically. He had read Newton's *Principia* and had mastered calculus, enough to use it in designing a new and innovative kind of plow. He had been a county surveyor in 1773 in his native Virginia; though a political appointment, the post reflected his abilities. Politically, he was deeply involved in the events of the American Revolution, being the principal author of the Declaration of Independence in 1776 and governor of Virginia in 1779. His famous *Notes on the State of Virginia* was the first systematic and comprehensive study of a region of the new country. From 1785 to 1789 Jefferson was a minister to France (where, historians say, the teenage slave Sally Hemings probably first became Jefferson's chambermaid and mistress) and returned at the invitation of George Washington to be the nation's first secretary of state.

THOMAS JEFFERSON, WHO AS THE FIRST U.S. SECRETARY OF STATE (1789–1793) WROTE THE COUNTRY'S FIRST PLAN TO ESTABLISH "UNIFORMITY FOR CURRENCY, WEIGHTS, AND MEASURES"—WHICH CONGRESS FAILED TO APPROVE—AND WHO AS THE COUNTRY'S THIRD PRESIDENT (1801–1809) ESTABLISHED THE U.S. COAST SURVEY, THE FIRST FEDERAL INSTITUTION TO OVERSEE U.S. WEIGHTS AND MEASURES.

Jefferson received instructions to work on a measures proposal on April 15, 1790, and finished a draft about May 20. He proposed a decimal system whose standard was tied to a seconds pendulum at the median U.S. latitude of 38 degrees. A few weeks later, he unexpectedly received a copy of Talleyrand's speech to the National Assembly and was impressed that Talleyrand fixed the latitude where the French seconds pendulum would be measured at 45 degrees, hoping to attract British participation. That latitude was then the country's principal northern boundary, the highest extremity of New York and Vermont, and thus inconvenient for U.S. scientists, but Jefferson went along, "with the hope that it may become a line of union with the rest of the world." He was thus willing to ally his revolutionary new plan for U.S. weights and measures with those brewing in Europe.

Jefferson sent his proposal to two trusted acquaintances for critique, Treasury Secretary Alexander Hamilton and David Rittenhouse, a noted Philadelphia astronomer and instrument maker and the previous president of the American Philosophical Society. Hamilton liked the report and had no substantive comments. Rittenhouse was lukewarm and commented extensively. Half a century before the British discovered the same thing, Rittenhouse realized that the pendulum would not make a good standard, and he outlined to Jefferson its numerous sources of error. Rittenhouse counseled his friend Jefferson that the traditional approach—making an artifact standard—would be superior. But Jefferson was too enamored with the ingenuity of this natural standard idea to abandon it. Learning that a Philadelphia clockmaker built pendulums of rods rather than wires, Jefferson included that idea, hoping that it would overcome Rittenhouse's objections. Jefferson completed his "Plan for Establishing Uniformity in the Coinage, Weights, and Measures of the United States" on July 4, 1790, submitting it to the House of Representatives on July 13. It proposed a standard pendulum equipped with a cylindrical rod made of iron and used at sea level. Jefferson wrote that any inaccuracies in such a device, already small now, will become less and less given the tendency of science to "progress towards perfection."

Jefferson pointed out that the Continental Congress had already abolished the British system of coinage, based on pounds, shillings, pence, and farthings, and substituted a decimal coinage based on dollars and cents. The House of Representatives might consider extending "a like improvement" to the new weights and measures system. Such an improvement would be "soon and sensibly felt by the whole mass of the people," who would be able to compute purchases much more readily than with "the present complicated and difficult ratios," encouraging equality among the citizenry. However, Jefferson also appreciated the difficulty that "changing the established habits of a whole nation" might be "an insuperable bar to this improvement." He therefore proposed two plans to the House of Representatives, one involving decimal, the other traditional units. Both proposals tied the units to a seconds pendulum.

PROPOSAL 1. The first proposal was to keep the present weights and measures inherited from the British, but make these "uniform and invariable" by tethering these to a natural standard. He appealed to the Carysfort Committee of 1758 and 1759 as "the best written testimony existing of the standard measures and weights of England." He divided the standard rod of 45 degrees into 587½ equal parts, defined this as a "line," then matched this to the linear British measures as follows:

10 lines an inch	5½ yards a perch or pole
12 inches a foot	40 poles or perches a furlong
3 feet a yard	8 furlongs a mile
3 feet 9 inches an ell	3 miles a league
6 feet a fathom	

Jefferson proposed to keep the area measures (an acre of 4 roods, a rood of 40 square poles) the same. For capacity, a gallon would consist of 270 cubic inches. Its subdivisions would include the quart (4 in a gallon) and the pint (2 in a quart). Its multiples would include the peck (2 gallons); the bushel or firkin (8 gallons); the strike or kinderkin (2 bushels); the barrel or coomb (2 strikes); the hogshead or quarter (2 coombs); the pipe, butt, or puncheon (2 hogsheads); and the ton (2 pipes). For weight, Jefferson proposed that an ounce would be the weight of a cube of rainwater one-tenth of a foot on each side—or, equivalently, a thousandth the weight of a cubic foot of rainwater. Its subdivisions would be the pennyweight (18 in an ounce) and the grain (24 in a pennyweight); its multiples would be the pound (16 ounces).

PROPOSAL 2. Jefferson's second proposal was for a "thorough reformation" of the entire system. In this, a seconds pendulum would be divided into five equal parts, each of which would be called a "foot." This foot would be divided into 10 inches, inches into 10 lines, and lines into 10 points. Further, 10 feet would make a decad; 10 decads, 1 rood; 10 roods, 1 furlong; and 10 furlongs, a mile. Area measures would be mainly in squares of these measures; measures of capacity would be in cubic feet, or bushels. Bushels would be subdivided into 10 pottles, pottles into 10 demi-pints, demi-pints into 10 meters—each meter a cubic inch. Multiples of the bushel would be the quarter (10 bushels) and the double ton (10 quarters). The standard weight measure would be the weight of a cubic inch of rainwater, defined as the ounce. This would be subdivided into double scruples (10 per ounce), carats (10 in a double scruple), minims or demi-grains (10 per carat) and mites (10 per minim); ounces would be multiplied into the pound (10 ounces), the stone (10 pounds), kental (16 stones), and hogsheads (10 kentals).

Jefferson later heard that the French were shifting from the idea of a pendulum to that of a meridian. He was disappointed; the shift seemed to betray the ambitious attempt toward universality that had been one of the chief merits. He wrote:

> The element of measure adopted by the National Assembly excludes, *ipso facto*, every nation on earth from a communion of measure with them; for they acknowledge themselves, that a due proportion for admeasurement of a meridian crossing the forty-fifth degree of latitude, and terminating at both ends in the same level, can be found in no other country on earth but theirs. It would follow then, that other nationals

> must trust to their admeasurement, or send persons into their country to make it themselves, not only in the first instance, but whenever afterwards they may wish to verify their measures. Instead of concurring, then, in a measure which, like the pendulum, may be found in every point of the forty-fifth degree, and through both hemispheres, and consequently in all countries of the earth lying under that parallel, either northern or southern, they adopt one which can be found but in a single point of the northern parallel, and consequently only in one country, and that country is theirs.[11]

Jefferson's report arrived on July 13, 1790, at a critical moment for U.S. measures reform. Western lands were being seized, settled, and surveyed; any delay implementing a new system would make it harder to overturn the existing one. But Congress put off discussion, adjourned on August 12, and returned only in December. On Tuesday, December 7, in Washington's second message to Congress, Washington again urged action on weights and measures, and Jefferson's report was assigned to a committee.

But the U.S. Congress never acted on Jefferson's ambitious system, distracted as it was by other pressing concerns. Washington again urged Congress to act in his third address in October 1791. The Senate appointed a committee, which in April 1792 approved Jefferson's proposals but postponed asking the Senate to consider it. More delays followed; no action. Congressional committees took up the idea again in 1798, 1804, and 1808. Still no action.

Jefferson was elected the nation's third president in 1801 and managed to have a far-reaching effect on American weights and measures by another route: his interest in surveying. In 1803, he negotiated the Louisiana Purchase, a huge amount of territory from France that almost doubled the territory of the young nation. He was interested in surveying not only that territory but other lands as well, dispatching the Louis and Clark expedition to explore a route to the Pacific Coast and its resources. High on Jefferson's agenda was to find the means to survey the United States on a large scale.

A solution arrived in 1806, via a letter of introduction from Robert Patterson, Jefferson's director of the mint, concerning a certain Ferdinand Hassler, a recent Swiss immigrant:

> He is a man of science & education; and . . . a character of considerable importance in his own country. It is his wish to obtain some employment from the United States, which would require the practice of surveying or astronomy. He would willingly engage in an exploring expedition, such as those you have already set on foot.[12]

Patterson said he'd enclosed a short biography that he'd asked Hassler to compose, then mentioned other qualifications:

> Besides his knowledge of the Latin language, he speaks the German, French, Italian & English. To his acquaintance with mathematics in general, which, as far as I am capable of judging from a short though not slight acquaintance, is very extensive, he adds a good knowledge of chemistry, mineralogy, and all the other branches of natural philosophy. In short, Sir, I believe his services may be rendered useful to this his adopted country. He possesses a very valuable library, and a set of surveying & astronomical instruments, scarce inferior to any I ever saw.

Hassler (1770–1843) also proved difficult to collaborate with. He was "egotistical, exasperating, and uncooperative," in the words of a modern historian of science.[13] But he became an important figure in early U.S. surveying as well as in establishing U.S. weights and measures.

Ferdinand Rudolph Hassler,
first superintendent of the U.S. Coast Survey.

Hassler had learned an appreciation for instruments from his father, a Swiss clock manufacturer, and was soon carrying out fieldwork on geodesy, mapping Bern. Fighting in Europe led him to come to the United States to become a farmer. He brought some books and instruments along, including a copy of the Committee meter and the Committee kilogram, as well as a standard French toise and English pound.

In Philadelphia, Hassler met several friends of Jefferson's, including the Director of the Mint Patterson; John Vaughan, a wine merchant who was librarian of the American Philosophical Society; and Arthur Gallatin, an urbane Swiss-born diplomat, administrator, and public servant. All were impressed by Hassler's knowledge of weights and measures and by his surveying ability, and realized that Hassler's skills could be useful to the young nation. Vaughan purchased Hassler's weights and measures, and wrote to Jefferson: "The important object of an universal Standard of Weight & Measure has long occupied your thoughts; you will therefore learn with pleasure that I have lately become possessed by purchase from Mr. Hassler—of the French Standard Toise-metre,-Kylogram & the English Troy Standard—Which may serve as standards of Comparison, whenever the subject is taken up in this Country."[14]

Jefferson, initially skeptical, soon recognized Hassler's skills. He proposed that Congress authorize a systematic survey of the coast of the United States, which was approved in 1807. Hassler was nominated to be its superintendent and dispatched to Europe to procure surveying instruments and more weights and measures. But the project ran into delays, Hassler's return was interrupted by the War of 1812, and only in 1816 did Hassler officially assume the position of superintendent of the U.S. Coast Survey. His tenure did not last long. He worked slowly and alienated people with his superior manner. Furthermore, many in Congress did not approve of a foreigner being in charge of such a prestigious project, and Hassler's Congressional opponents passed a bill mandating that the Survey be led by U.S. military personnel. Hassler was forced to step down, though he would eventually regain the post.

That year, 1817, the U.S. Senate asked Secretary of State John Quincy Adams to prepare a report on the weights and measures used abroad, the different regulations of the states, and the prospects for establishing uniform weights and measures in the United States. Adams (1767–1848) had spent much time abroad, first accompanying his father John Adams (1735–1826), who was an envoy to France and the Netherlands before becoming vice president and president; and then on trips to Finland, Sweden, Denmark, Silesia, and Prussia; and finally as an ambassador to Russia and then Britain. During these trips Adams was fascinated by the still-diverse European systems of weights and measures and with the new metric system. In 1817 President James Monroe called Adams back from Britain to become secretary of state, whereupon he got handed the long-delayed task of reopening the weights and measures question.

Adams was a workaholic during a demanding time in American

history. Among other things, he wrote the Monroe Doctrine, which announced that the United States would resist European incursion into the Americas and avoid interfering with other states, famously declaring that America "goes not abroad in search of monsters to destroy." He found researching weights and measures as compelling and spent three years on it. He asked each state for its laws and systematically reviewed European systems. He used no interns or assistants to research the document, nor ghostwriters to draft it, but researched and wrote it himself, rising at 5 a.m. and sometimes earlier to write by candlelight. He rejected his father's suggestion that he put down the project to take the annual family vacation in Maryland. His wife complained: "His whole mind is so intent on weights and measures that you would suppose his very existence depended on this subject."[15]

Adams's ambitions for the report grew. "Where the project had first promised to be merely a dry enumeration of tables and formulas," wrote one historian, "under JQA's powers of interpretation, it had become a vision of what government could do for public well-being."[16]

In 1821, Adams submitted to Congress a 135-paged report, with an additional hundred pages of appendices, covering the history of weights and measures and analyzing the prospects and problems of reform. It addresses theory and practice, scientific and political issues, and philosophical and moral questions. It is not easy to read, especially now in the PowerPoint age: there are no helpful bullet points, outlines, or summaries. It is repetitive, and often moves from minutia to eloquence in just a few sentences. Yet it is far-ranging and thoughtful.

Adams begins by tracing the origins of weights and measures from bodily measures to the need to multiply as well as standardize weights and measures after the onset of civil society. He is sensitive to the issues of trust and morality raised by systems of weights and measures, and the difficulties of legislation: "nature has planted sources of diversity, which the legislator would in vain overlook, which he would in vain attempt to control." In a thorough comparison of the British and French systems, Adams vastly preferred the French and declared the British to be "the ruins of a system," its nomenclature "full of confusion and absurdity." The French system, if enacted universally, would be a great step forward for humankind. "This system approaches to the ideal perfection of uniformity applied to weights and measures; and, whether destined to succeed, or doomed to fail, would shed unfading glory upon the age in which it was conceived, and upon the nation by which its execution was attempted, and has been in part achieved." This system of measures, Adams hoped, would forge bonds "between the inhabitants of the most distant regions," surrounding the globe, and "one language of weights and measures will be spoken from the equator to the poles."[17] He stresses:

> Weights and measures may be ranked among the necessaries of life to every individual of human society. They enter into the economical arrangements and daily concerns of every family. They are necessary to every occupation of human industry; to the distribution and security of every species of property; to every transaction of trade and commerce; to the labors of the husbandman; to the ingenuity of the artificer; to the studies of the philosopher; to the researches of the antiquarian; to the navigation of the mariner; and the marches of the solider; to all the exchanges of peace, and all the operations of war. The knowledge of them, as in established use, is among the first elements of education, and is often learned by those who learn nothing else, not even to read and write. This knowledge is riveted in the memory by the habitual application of it to the employments of men throughout life.[18]

Adams considered the metric system "a new power, offered to man, incomparably greater than that which he has acquired by the new agency which he has given to steam. It is in design the greatest *invention* of human ingenuity since that of printing." Additionally, it has moral advantages in the way it facilitates and simplifies transactions among human beings, encouraging equality, and Adams compares it to moral and political improvements such as eliminating slavery. A superior measurement system gives nobody any greater advantage, it masks "no project of avarice or ambition," it disguises "no private or perverted ends." How odd it would be, Adams wrote, if human beings could fashion identical weapons with which to destroy each other, but be unable to eat and drink with the same weights and measures!

But Adams also foresaw the inevitable legislative difficulty. Weights and measures are woven into every aspect of life. To change them would "affect the well-being of every man, woman, and child." Adams thought the United States would face even stronger obstacles than France was encountering, even with that country no longer in revolution. At the end of his document, Adams recommended no change in the existing weights and measures of the United States, but did recommend that Congress "consult with foreign nations, for the future and ultimate establishment of universal and permanent uniformity." If uniformity of weights and measures could be enacted, it would be "a blessing of such transcendent magnitude," that those who accomplished it "would be among the greatest of benefactors of the human race." But we are not there yet. When it will extend beyond France, he concluded, "must await the time when the example of its benefits, long and practically enjoyed, shall acquire that ascendancy over the opinions of other nations which gives motion to the springs and direction to the wheels of power."[19]

"No other philosopher or political economist in the world," writes the historian William Appleman Williams, "ever personalized and humanized the elementary problem of weights and measures—or any other mundane but vital element of their system—in a comparable manner."[20]

Adams's *Report* is a wise piece of writing that is too long because its author did not have the time to write shorter. Few people read it all the way through, not even his father. Congressmen, glancing at the conclusion, decided it was OK to do nothing. That year, 1821, Gallatin—then the minister to France—sent back a platinum kilogram and meter. Six years later, now minister to London, Gallatin brought back to the now President Adams an imperial pound weight, a copy of the 1758 weight that was certified by the Act of 1824. It was received by Adams in a special ceremony on October 12, 1827. Enclosed in a cask sealed in London, it was used as a U.S. standard for almost 70 years. The next year, 1828, a particular brass troy pound housed at the Philadelphia Mint was designated as the "standard troy pound" for the purposes of coinage.

In 1832, Congress took some action by mandating a comparison of the weights and measures in the country's custom houses. Hassler, meanwhile, was rehabilitated as superintendent of the Coast Survey and given the task. In the course of this work, Hassler produced the first official scientific document of the U.S. Office of Weights and Measures, and therefore one of the first official scientific documents issued by the U.S. government.[21] In 1836, the treasury secretary was assigned the task of creating and distributing a set of weights and measures to the governor of each state. That was as far as Congress had gone toward fixing standards of weights and measures. No system was formally instituted—neither the imperial system nor the metric—and, except for special purposes such as customs and the mint, the states were left to decide matters on their own.

FRANCE

On November 9, 1799—a famous date known as 18th Brumaire in Year VIII of the Revolutionary calendar—a general named Napoleon Bonaparte staged a coup and overthrew the Directory. Rumors flew that the new metric system, a product of revolutionary fervor, was dead. A year later, Napoleon began to water down the system by reintroducing new names alongside the old; the *livre métrique*, for instance, would be equal to the kilogram. Six years later, Napoleon discarded the revolutionary calendar, though he left the newly instated metric system and the old system to coexist uneasily. In 1812, in yet another retreat from the metric system, he allowed nonmetric fractions and multiples to resurface. French enthusiasm for rationality and universality in their weights and measures appeared to be on the wane. As the French politician Benjamin Constant put it:

> One code of laws for all, one system of measures, one set of regulations . . . this is how we perceive today the perfection of social organization. . . . uniformity is the great slogan. A pity, indeed, that it is not possible to raze all towns to the ground so as to be able to rebuild them on one and the same pattern, and to level mountains everywhere to a single preordained plain. Indeed, I am surprised at the absence, as yet, of a ukase ordering everybody to wear identical clothes, so that the sight of the Lord will not encounter lack of order and offensive diversity.[22]

In 1814, Napoleon himself was deposed, and rumors again flew that the metric system was dead. But the new regime reaffirmed the system, though again without abandoning the old, for another quarter century. In Adams's report of 1821, he noted that the old names were still in use, which surprised him. The nomenclature of the metric system, he said, is "perfectly simple and beautiful":

> Twelve new words, five of which denote the things, and seven the numbers, include the whole system of metrology; give distinct and signifi-

cant names to every weight, measure, multiple, and subdivision, of the whole system; discard the worst of all the sources of error and confusion in weights and measures, the application of the same name to different things; and keep constantly present to the mind the principle of decimal arithmetic, which combines all the weights and measures, the proportion of each weight or measure with all its multiples and divisions, and the chain of uniformity which connects together the profoundest researches of science with the most accomplished labors of art and the daily occupations and wants of domestic life in all classes and conditions of society.[23]

Yet the revolutionary French system and its nomenclature have had trouble sticking, Adams noted. The French "have refused to learn, or to repeat these twelve words. They have been willing to take a total and radical change of things; but they insist upon calling them by old names. They take the metre; but they must call one-third part of it a foot. They accept the kilogramme; but, instead of pronouncing its name, they choose to call one half of it a pound."

Additional problems emerged with the metric system. Errors were found in Méchain and Delambre's work, as well as in the calculations of the meridian, meaning that the Meter of the Archives—the étalon—was a few lines (about 0.2 mm) shorter than the definition of one ten-millionth of the Paris meridian. Not only that, the Kilogram of the Archives was a little lighter than a cubic decimeter of water—and fashioning a vessel of exactly a cubic decimeter was much more difficult than envisioned. Still, the scientists decided to use the standards to define the units for two reasons: changing them could be terribly inconvenient, and it made no practical difference.

During a scientific gathering in Paris in 1827, some scientists pointed out that if a comet or asteroid struck Earth and altered its shape and axis of rotation, this could change both the meridian measurement and the seconds pendulum. How would humankind define the meter then? Earth, evidently, might not be as good a source for natural standards as scientists of the previous century had assumed. Ingenious members of the group began dreaming up workable standards that would be independent of the earth's dimensions. The British chemist Sir Humphry Davy (1778–1829) proposed using capillary action, the phenomenon that liquid rises in confined spaces due to interatomic forces; he proposed that the basic length measure could be the diameter of a thin tube of glass which sucked up water by the same amount as that diameter, which he presumed to be a unique length independent of gravity. Realizing the difficulties of putting Davy's definition into practice, the French physicist Jacques Babinet (1794–1872), who was developing measures for wavelengths of light, suggested using light wavelengths as the basic length standard instead. "Although these two projects are far from new and the techniques for observing them might have been put into place, no one has done so," Babinet wrote. "And, truth be told, there is little to regret in this, for it must be recognized that they are of no real utility."[24]

Louis Philippe (1773–1850), who became the French King in 1830, had the metric system reviewed in 1837 and restored. By a decree of July 4, 1837, the metric system was mandated to take effect on January 1, 1840. After that date, anyone found violating the law by using nonmetric units would be fined ten francs for every nonmetric measurement. The general adoption of the metric system therefore took nearly half a century in the land of its birth.

• • •

From 1790 to 1850, France, Great Britain, and the United States had pursued very different paths toward new systems of weights and measures. The French had mandated the metric system for only 10 years, following half a century of struggle. In the process, they had found that their idea of attaching the meter to a fraction of the earth's meridian was impractical. The British had consolidated their weights and measures into the imperial system, but had found that their idea of attaching the yard to the length of a seconds pendulum was impractical. The United States had yet to fix a standard of weights and measures, or even to declare any system to be legal. The metric system had yet to become universal, and the dreams of a natural standard that had inspired it had been shattered.

CHAPTER SIX

"ONE OF THE GREATEST TRIUMPHS OF MODERN CIVILIZATION"

By the mid-nineteenth century, the British Empire looked, and felt, unstoppable. On May 1, 1851, it threw itself a party. All London, it seemed, came out to Hyde Park to celebrate the opening of the Great Exhibition of the Works of Industry of All Nations, the first exhibition of its kind. Queen Victoria herself presided over the opening ceremonies, leaving for the park with Prince Albert and her two eldest children shortly after eleven that morning, she dressed in pink watered silk and wearing the Koh-i-noor diamond, her husband in uniform. As her parade of nine state carriages left Buckingham Palace, she was met with a happy, cheering crowd as far as the eye could see. The gleaming glass exhibition hall slowly emerged into view. Almost 2000 feet long, it was made of a million square feet (300,000 panes) of glass framed between thousands of cast iron girders. The enormous structure was designed to house 100,000 exhibitions by thousands of exhibitors, and its 64-foot-high central dome was gaily decorated with the flags of the exhibition's thirty-four participating nations.

When Victoria's carriage stopped in front of the "Crystal Palace," as it had been dubbed by the initially mocking *Punch* magazine, Victoria was escorted inside by Albert and their children, past colorful banners and drapes, exotic plants collected from all over the world, and fully grown elm trees, to the central nave. There she took her seat on a throne while Albert gave an official opening speech: "Science discovers these laws of power, motion, and transformation; industry applies them to the raw matter which the earth yields us in abundance, but which becomes valuable only by knowledge. Art teaches us the immutable laws of beauty and symmetry, and gives to our productions forms in accordance with them."[1]

After a prayer by the Archbishop of Canterbury, a performance of Handel's Hallelujah Chorus by a choir of 600, and gun salutes, the royal couple toured the galleries, and the exhibition was opened to the public.

THE GREAT EXHIBITION OF 1851

The exhibition, though not Prince Albert's idea, had become his passion. Small-scale displays of manufacturing and engineering had been staged in various cities for a decade or so. Albert's ambitions were larger. He wanted to showcase Britain's manufacturing and engineering prowess, and the prosperity and world dominance that this had brought the nation, but also the worldwide Industrial Revolution—its beneficial impact on civilization, its elegance, and even its beauty. The Great Exhibition was the opening of a new era, sometimes called the "second" Industrial Revolution, a period of rapidly increasing mechanization and international trade and collaboration.

Inadvertently, the exhibition showcased something else, as well—that the second Industrial Revolution was in danger of being cramped by poorly coordinated measurement systems even in England. The exhibition set in motion efforts to reform these measurement systems, efforts that would culminate, almost a quarter century later, in an international treaty to create an international agency to supervise the world's weights and measures.

In the first half of the nineteenth century, the metric system had been established in France after prolonged birth pangs and was accepted by only four other countries, three of which were neighbors of France—Belgium, tiny Luxembourg, and the Netherlands—and Algeria, a French colony. Resistance from other countries sprang from lingering fears of the excesses of revolutionary France and lack of serious discontent with existing systems. When the metric system was finally mandated in France in 1840, and the French Foreign Minister François Guizot sent standards to several different countries trying to promote the system, he got a lackluster response.

The pace of international acceptance was about to speed up. The Great Exhibition's international collection of machinery highlighted the need for greater engineering precision and international cooperation in setting measurement units and standards. On exhibit were several innovative measuring machines, including devices that measured to a millionth of an inch built by British inventor Joseph Whitworth. Yet in trying to compare the machines and instruments, judges found themselves hampered by the different measures used by exhibitors of different countries. Among the exhibitors, however, was the French Conservatoire des Arts et Métiers, whose exhibit included metric measures. The exhibition attracted attention both to the problem and to a promising solution.

The judges' difficulties prompted the British Society for the Encouragement of Arts, Commerce, and Manufactures to recommend a decimal system of measures, "a most important step in advancing the Arts, Manufactures, and Commerce of the country," and indeed for the "adoption of a uniform system throughout the world," with the metric system being the outstanding candidate.[2]

One early promoter of international collaboration, William Farr (1807–1883), came from the medical field. A British medical statistician, Farr had firsthand experience with the perils of nonuniform terms:

> The advantages of a uniform statistical nomenclature, however imperfect, are so obvious, that it is surprising no attention has been paid to its enforcement in Bills of Mortality. Each disease has, in many instances, been denoted by three or four terms, and each term has been applied to as many different diseases: vague, inconvenient names have been employed, or complications have been registered instead of primary diseases. The nomenclature is of as much importance in this department of inquiry as weights and measures in the physical sciences, and should be settled without delay.[3]

The first International Statistical Congress (ISC), which met in Brussels in 1853, passed a resolution standing that "in the statistical tables published in the countries where the Metric system does not exist, there should be added a column indicating the metrical reduction of weights and measures."[4] At the second ISC in 1855, participants formed an organization to promote international adoption of a universal system of weights and measures, the International Association for Obtaining a Uniform Decimal System of Measures, Weights, and Coins. The organization's first President, Baron Rothschild, the famous scion of the renowned Rothschild banking family of France, threw his support behind the metric system. Participants at the fourth ISC in London in 1860 called for members to champion the system in their home countries. By then, four more countries had added themselves to the list of those adopting the metric system: Colombia, Monaco, Cuba, and Spain.

Successive international exhibitions in the next two decades, including ones in Paris in 1855 and London in 1862, did still more to advance the metric cause. Even in Britain, the initial signs looked promising. The House of Commons appointed a committee to conduct hearings about the issue. The committee found few supporters of the imperial system and recommended that Parliament begin "cautiously but steadily to introduce the metric system into this country."[5] But delays and lack of interest in the House of Lords watered down the bill, and the resulting Weights and Measures Act of 1864 allowed metric units in British contracts but did not legalize it in trade. In 1868, another Royal Commission recommended a stronger bill requiring adoption of the metric system. Again the measure passed the House of Commons and died in the House of Lords.

The loss of the imperial standards in the dramatic burning of Parliament, and the inability to re-create them using a seconds pendulum, had shaken the confidence of many eminent British scientists in natural standards. "[O]ur yard," said Herschel, who had served on the committee to recover the standards, "is a purely individual material object, multiplied and perpetuated by careful copying; and from which all reference to a natural origin is studiously excluded, as much as if it had dropped from the clouds."[6] The lesson seemed to be that standards were fated to be arbitrary. The purpose of a country's measurement standards is to make its industries flourish: nothing else matters. By that standard, Britain's metrology was unsurpassed. Herschel wrote, "Taking commerce, population and area of soil then into account, there would seem to be far better reason for our continental neighbours to conform to *our* linear unit . . . than for the move to come from our side."[7]

A year after the defeat of the 1868 metric bill in Great Britain, the British Standards Commission went out of its way to dampen enthusiasm for the metric system, citing the long and illustrious history of imperial standards in British history. Patriotism alone seemed a sufficient reason to retain imperial standards.

The social Darwinist thinker Herbert Spencer, who coined the phrase "survival of the fittest," even invoked philosophical issues in his opposition to the metric system. He roundly criticized scientists for preferring the elegance of a unified system and considered the idea of a permanent standard to be an abomination and even against nature, which thrived on diversity and cut-throat conflict. When he died in 1903, his will contained a provision to the effect that, whenever a pro-metric bill was introduced into the British Parliament, his remarks on the subject should be reprinted and distributed to its members.

Though Great Britain failed to adopt the metric system, the fact that this great industrial power almost did so was worldwide news and increased the metric system's reputation. In the United States, the National Academy of Sciences took up the cause in 1863, and recommended its adoption to Congress. In 1866, the U.S. Congress enacted the following law, signed by President Andrew Johnson:

> AN ACT to authorize the use of the metric system of weights and measures. *Be it enacted by the Senate and House of Representatives of the United States of America in Congress assembled*, That from and after the passage of this Act it shall be lawful throughout the United States of America to employ the weights and measures of the metric system, and no contract or dealing, or pleading in any court, shall be deemed invalid or liable to objection because the weights or measures expressed or referred to therein are weights or measures of the metric system.[8]

Like the British Act of 1864, Jackson's law hardly changed anything, officially at least. It did not declare uniform standards but only made the metric system legally acceptable. Yet it was the first piece of general legislation in the United States on the subject of weights and measures, and the first to declare a system—any system—legal throughout the land.

In so doing it directed attention to the metric system. The next several decades saw a spurt of pro-metric activity. Several states, including Connecticut, New Jersey, and Massachusetts, encouraged its teaching in schools. The 1866 Act also provided the legal framework that allowed the United States to participate in international negotiations involving the metric treaty, which began shortly.

These negotiations were set in motion in 1867—a key year for pro-metric advocacy—in the wake of more international congresses. At the sixth ISC (Florence, 1867), participants urged "universal adoption" of the metric system and asked members to form advocacy groups. At the Paris International Exposition of 1867, a special circular pavilion in the center of its central garden displayed weights, measures, and coins of participating nations, as well as an extensive display of France's pride, the metric system. This conference was headed by the emperor Louis Napoleon III, Bonaparte's nephew and eventual successor.

The 1867 meeting of the International Geodetic Association—which had been the first of the burgeoning number of international scientific organizations—had a far-reaching impact on the metric system. Geodesy is the study of the exact shape of the earth, a field that had grown in importance ever since Newton had proposed that the earth was not a simple sphere but—because of centrifugal force—slightly flattened at the poles: an "oblate spheroid," or soccer ball squeezed at the top and bottom. Could this be true? Early in the eighteenth century, measurements with seconds pendulums had clearly demonstrated this to be the case, and the question became whether other variations existed in the earth's shape. The diversity of measures in different countries complicated the quest to answer this question. The international interest in geodesy now drove the effort to universalize the metric system.

In the United States, still in the process of acquiring territories, geodesy was an increasingly important concern of the Coast Survey. By 1843, when its first superintendent, Julius Hassler, died and was replaced by

Alexander Bache, the Survey had established a base line on the south coast of Long Island and extended triangulation northwest to Rhode Island and southwest to Chesapeake Bay, covering about 9000 square miles of territory. Under Bache the Survey further expanded its surveying, as the United States had acquired more territory in Texas and areas further west. The Survey was now the preeminent scientific agency of the U.S. government, employing scientists and conducting research in the national interest. When Bache died in 1867, he was succeeded by Harvard mathematician Benjamin Peirce. In the 1860s, under Peirce, the agency's political profile continued to climb because of the agency's work connected with the purchase of Alaska from Russia in 1867 and with the proposed purchase of Greenland and Iceland from Denmark. But the Survey's chief focus would soon shift from surveying to geodesy.

During the mid-nineteenth century, the practice of geodesy had two implications for metrology. One was the need for an international network of collaborating scientists; geodesy only made sense as a global project. The other was increased need for precision measurements of gravitational variations, known as gravimetrics.

The first two meetings of the International Geodetic Association were held in Berlin in 1864 and 1867. Participants at the second meeting strongly urged construction of metric standards with new precision. But they went further than their colleagues in other fields by urging construction of a new standard meter, to replace the Meter of the Archives. In the interim years since the first meter and kilogram standards had been constructed, advancing technology (alloys sturdier than platinum) made it possible to construct more reliable standards. Finally, the conference participants made a far-reaching suggestion: an international commission of scientists should be charged with overseeing construction of this new standard and maintaining it once built. This was clearly in the international spirit of the system itself but would take the metric system out of the hands of France.

French scientists initially balked. But on September 2, 1869, Emperor Napoleon appointed an International Commission for the Meter, to meet in Paris in August 1870. Twenty-five foreign countries accepted an invitation to attend, including the United States, Great Britain, and Russia.[9] By this time, too, eight more countries had adopted the metric system: Brazil, Mexico, Italy, Uruguay, Chile, Ecuador, Peru, and Puerto Rico. At last the metric system seemed on the verge of international acceptance.

"The battle of the standards is over," announced *Nature*, the English-speaking-world's leading science magazine, in 1870, commenting on the impending gathering in Paris, "and we may say the Metre has gained the victory." The imperial system, though still in use, was doomed to obsolescence. But the metric system had not prevailed because it was tied to a universal and natural standard; *Nature* even made fun of the view that a "ten millionth part of the quadrant of the earth" could come anywhere close to being measurable with precision. Rather, the meter won out because it "is already a cosmopolitan unit, widely recognized, and in general use among many nations; and that whilst other units remain as philosophical abstractions, the metre is the basis of a system, not only perfectly complete, homogeneous, and scientific, but simple and practical in all its parts."[10] In short, the meter was universal because it was universal. All that remained to clean up British metrology, *Nature* continued, was a propitious moment to mandate it. The wait would take longer than *Nature* anticipated.

THE INTERNATIONAL METRIC COMMISSION

Metric advocates, already accustomed to delays, experienced another. The historic International Metric Commission commenced on August 8, 1870. But just 3 weeks previously, war had broken out between France and Prussia, and on August 4 the combined armies of Prussia and several German states crossed the border into Alsace, quickly defeated the French army, and headed for Paris. The Commission hastily adjourned, planning to meet again when they could.

During the year-long war, in which Paris was under siege for 4 months, French scientists investigated materials for the new standards. They decided against using any existing metals, which either corrode or vary in density. A stone like quartz was solid and did not corrode, but was fragile. Glass attracted water condensate and expanded or shrank in size with temperature. The platinum of the previous standard was soft and weak. The French delegates decided on a new platinum-iridium alloy, 90 percent platinum and 10 percent iridium, for both the meter and the kilogram standard. The new meter would have a peculiar cross-section, somewhat X-shaped but with a bar across the middle, slightly lower than dead center. The new meter rod would also be a "line" rather than an "end" standard; it would be slightly longer than a meter, with defining lines cut a centimeter from each end. It would be supported at points called Airy points, which the dedicated British scientist had calculated would involve the least bending and drooping of the standard.

After the war, when France was stable again and traveling safe, the International Metric Commission was rescheduled for 1872. This time, representatives of thirty countries attended. Julius Hilgard (1825–1891) was the U.S. delegate. Born in Germany, Hilgard's family emigrated to the United States in 1836, and he studied engineering in Philadelphia. He entered the Coast Survey in 1844 and remained in its service for 40 years. Hilgard expected to succeed Bache as superintendent after his death in 1867, but dutifully continued to serve as an assistant and manager of the weights and measures office, and representative to foreign meetings.

Sa Majesté l'Empereur d'Autriche-Hongrie, Sa Majesté l'Empereur d'Allemagne, Sa Majesté le Roi des Belges, Sa Majesté l'Empereur du Brésil, Son Excellence le Président de la Confédération Argentine, Sa Majesté le Roi de Danemark, Sa Majesté le Roi d'Espagne, Son Excellence le Président des États-Unis d'Amérique, Son Excellence le Président de la République Française, Sa Majesté le Roi d'Italie, Son Excellence le Président de la République du Pérou, Sa Majesté le Roi de Portugal et des Algarves, Sa Majesté l'Empereur de toutes les Russies, Sa Majesté le Roi de Suède et de Norvège, Son Excellence le Président de la Confédération Suisse, Sa Majesté l'Empereur des Ottomans, et Son Excellence le Président de la République de Vénézuela, désirant assurer l'unification internationale

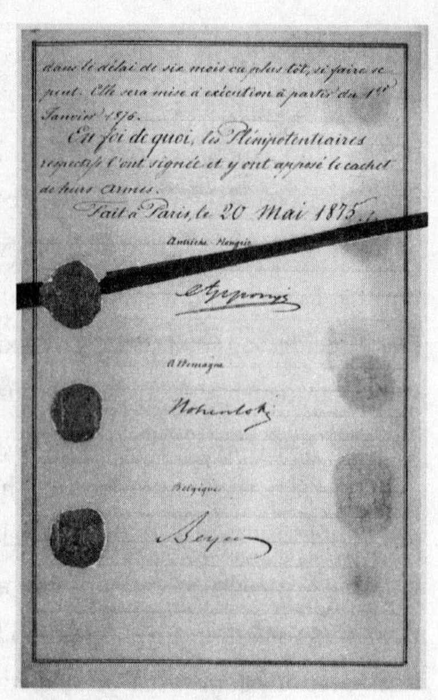

dans le délai de six mois ou plus tôt, si faire se peut. Elle sera mise à exécution à partir du 1er Janvier 1876.

En foi de quoi, les Plénipotentiaires respectifs l'ont signée et y ont apposé le cachet de leurs armes.

Fait à Paris, le 20 Mai 1875.

Autriche-Hongrie

Allemagne

Hohenlohe

Belgique

THE FIRST PAGE (LEFT) AND FIRST SIGNATURES (RIGHT) OF THE TREATY OF THE METER (1875), A LANDMARK IN THE HISTORY OF MEASUREMENT, INTERNATIONAL COOPERATION, AND GLOBALIZATION.

The delegates at the 1872 convention proposed to establish a truly international organization (at first called simply an International Metric Institution), the International Bureau of Weights and Measures, or BIPM, after its French acronym. It would be funded jointly by participating nations. Its responsibilities would include making and preserving the new standards, verifying those of other lands, and developing instrumentation. The BIPM was to be under the direction of an International Committee of Weights and Measures (CIPM), formed from all the delegates of the signing nations, whose members would meet in a General Conference of Weights and Measures (CGPM) every 6 years. A treaty was drafted outlining these ideas, and the delegates returned home to obtain permission to sign it at a conference to be held in 1875. If the proposed organization succeeded in establishing a "real and practical uniformity" in the weights and measures of the world, enthused *Nature*, it would be "one of the greatest triumphs of modern civilization."[11]

On May 20, 1875—a day now known as International Metrological Day—seventeen countries, including the United States, signed the treaty creating an International Metric Commission. It was a landmark in the history of measurement, international cooperation, and globalization. The treaty's terms gave up the idea of a natural standard—artifacts would have to do. That dream, scientists now felt, was far less important than worldwide agreement. The sentiment was captured by *Nature*: "[I]t ought also to be understood that for the present day the pith of the matter does not centre in the metre being a few microns (millionths of a millimeter) longer or shorter. The great point is that the whole world possess the same meter, and that the copies distributed be all perfectly equal to the standard, or rather rigorously determined in relation to the standard."[12]

Three countries did not sign: Great Britain, Holland, and Greece. All three were willing to support producing and maintaining new metric standards, but were leery of establishing a larger organization with more ambitious aims. Great Britain was further incensed by the fact that the General Conference was charged with "propagating" the metric system, thus potentially interfering with the imperial system.

France, meanwhile, gave the new international organization a small piece of land in the Parc Saint-Cloud, on the outskirts of Paris. André Le Nôtre, architect of the great gardens at Versailles, had designed the park but the buildings in it had been badly damaged during the Franco-Prussian War of a few years before. Though dilapidated, the Pavillon de Breteuil, which had once housed servants of the French king, was the Bureau's for the taking. The building was turned over on October 4,

Entrance to the International Bureau of Weights and Measures, outside Paris.

1875, and took a few years to fix up before the scientists could move in. When they did, in 1878, the BIPM became the world's first international laboratory.

In the 1870s, a dozen more countries adopted the metric system: Austria, Liechtenstein, Germany, Portugal, Norway, Czechoslovakia, Sweden, Switzerland, Hungary, Yugoslavia, Mauritius, and the Seychelles. Britain finally overcame its objections and signed the treaty in 1884. The editors of *Nature* were thrilled, believing that it heralded the imminent adoption of the metric system, not only by Britain but the entire world, and they reprinted an article from their French counterpart *La Nature* to the following effect:

> The universal introduction of a uniform system of weights and measures, by establishing a new bond between people and people and by promoting international relations, will undoubtedly prove a powerful factor in the interests of civilization. . . . More than anything else the interest of the labours of the bureau is scientific. Science will more and

more cease to rest content with close approximations; in all possible branches it craves rigorous exactitude, it aims at precision.[13]

In the meantime, the BIPM had had troubles manufacturing the new standards. The alloy failed to meet requirements and contained small amounts of impurities. To the embarrassment of the French, a London firm, Johnson Matthey & Company, was asked to manufacture the alloy. The company cast and recast the alloy several times to remove impurities, completing the work in 1884. In 1886–87, a French firm cut the material into lengths for the meter standard and fashioned cylinders for the kilogram standard. BIPM scientists chose a standard of each—the International Prototype Meter and the International Prototype Kilogram—as well as a set of *témoins*, or witnesses, for the use of the BIPM. Treaty signers would also receive a set.

In September 1889, the General Conference of Weights and Measures, the supervising agency of the BIPM, met for the first time. At this conference it accepted the standards as the official international standards and ratified what had been established years before—that, bowing to reality, the definition of the meter was not to be one ten-millionth of the earth's meridian running through Paris, but the length of this standard. A lottery determined distribution of the standards to the member countries. The United States received numbers 21 and 17 of the meter standard, and numbers 20 and 4 of the kilogram standard.

Benjamin A. Gould, the American representative to the International Committee who had succeeded Hilgard in 1887, packed and sealed the standards and gave them to the U.S. ambassador to Paris. The ambassador put them in the hands of a representative of the Coast Survey, who brought them to Washington, keeping them in sight and meticulously recording his route and their every location. When he took them on buses and railway cars he put them on cushioned seats. On November 15, 1889, he boarded the steamship *Germanic*, getting them a special stateroom.

On January 2, 1890, the crates arrived at the White House. President Benjamin Harrison broke the seals and opened the crates in the presence of Coast and Geodetic Survey superintendent Thomas C. Mendenhall and other dignitaries. The standards were then taken to a fireproof room in the Coast Survey building.

The Office of Weights and Measures had long struggled to make sure its standards and those of the British were separate but equal. Mendenhall saw an opportunity to make this unnecessary and on April 5, 1893, issued an order. The Constitution, he noted, gave Congress the power to "fix the standard of weights and measures," but that body had never actually exercised this power. In the absence of "material normal stan-

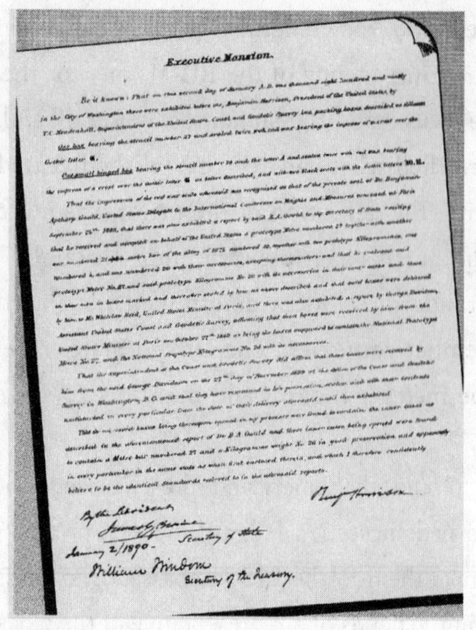
Executive Mansion.

By the President

January 2/1890 -

Secretary of State

Secretary of the Treasury.

PROCLAMATION OF JANUARY 2, 1890, ANNOUNCING RECEIPT BY TWENTY-THIRD U.S. PRESIDENT BENJAMIN HARRISON OF A SET OF OFFICIAL STANDARDS FROM THE INTERNATIONAL BUREAU OF WEIGHTS AND MEASURES.

dards of customary weights and measures," Mendenhall declared, "the Office of Weights and Measures, with the approval of the Secretary of the Treasury, will in the future regard the International Prototype Metre and Kilogramme as fundamental standards, and the customary units—the yard and the pound—will be derived therefrom in accordance with the Act of July 28, 1866." In practice, this had been the case for years; Mendenhall just wanted to make this formal "for the information of all interested in the science of metrology or in measurements of precision."[14] The official U.S. yard would be defined in terms of the meter, one yard is equal to 3600/3937 of a meter; the official U.S. pound would be derived from the kilogram by a ratio: 1 pound avoirdupois = 1/2.2046 kilograms. The Mendenhall order was an admission that, however long the United States might keep using imperial units, the French had won the battle of the standards.

Many scientists thought that the Mendenhall order heralded the imminent adoption in the United States of the metric system. The sooner the better, *Nature* wrote:

> It will be no easier for a hundred millions of people ten years hence to make the change than for seventy millions to-day. It is simply a question whether this generation shall accept the annoyance and inconvenience of the change largely for the benefit of the next, or shall the people of to-day selfishly consult only their own ease and impose on their children the double burden of learning and then discarding the present "brain-wasting system."[15]

The metric system had made great progress toward universality during the second half of the nineteenth century. At the century's midpoint, only Britain and British colonies or former colonies were using the imperial system while France and a handful of countries over which France had strong influence were using the metric system. In the second half of the nineteenth century, this changed, with the metric system becoming the measurement system of choice. Even Great Britain, in 1897, legalized use of the metric system in trade. Long before the establishment of the BIPM, the superiorities of the metric system over the imperial were clear. Both had accessible and assured standards, but the systematic character of the metric system, with its integrated and carefully scaled units, was much easier to use both in the laboratory and the marketplace. The imperial system had a hopeless array of measures without simple multiples or simple divisions, in which units were sometimes not even connected with each other via whole numbers:

SOME IMPERIAL SYSTEM UNITS

LENGTH	inch	VOLUME/CAPACITY	ounce
	hand (4 inches)		gill (5 fluid ounces)
	foot (12 inches)		pint (16 fluid ounces)
	yard (3 feet)		quart (2 pints)
	rod (16.5 feet)		gallon (4 quarts)
	chain (22 yards)		peck (2 gallons)
	furlong (40 rods)		bushel (4 pecks)
	mile (5280 feet)		
	league (3 miles)	AREA	perch (1 rod x 1 rod)
			rood (1 furlong x 1 rod)
MASS/WEIGHT	dram		acre (1 furlong x 1 chain)
	ounce (16 drams)		
	pound (16 ounces)		
	stone (14 ounces)		
	quarter (2 stone)		
	hundredweight (112 pounds)		
	ton (2240 pounds)		

The metric system, on the other hand, had simple and extendable multiples and divisions, solving the problem of appropriateness by using a common system of decimals and prefixes. You could go up and down the scale of dimensions easily, like a piano glissando:

METRIC SYSTEM UNITS AND PREFIXES
(NINETEENTH CENTURY)

UNITS	PREFIXES		
	mega (M)	1,000,000	10^6
	kilo (k)	1,000	10^3
	hecto (h)	100	10^2
	deca (da)	10	10^1
meter (m, length); gram (kg, mass/weight)		1	
	deci (d)	0.1	10^{-1}
	centi (c)	0.01	10^{-2}
	milli (m)	0.001	10^{-3}
	micro (μ)	0.000001	10^{-6}

Note: Area measures were made by squaring, volume/capacity measures by cubing length measures. Later, additional prefixes were added. The kilogram is peculiar in having a prefix yet also being the quantity defined by the standard, in that sense the base unit.

Yet habit and the cost of replacing an industrial infrastructure built with another system remained the chief obstacles to further expansion of the metric system. One could fill an entire book of examples of different local measurement systems from around the world, and the different ways in which countries using them were led to overcome these obstacles and replace them with the metric system. Oddly enough, the plunder, ravagement, and exploitation that accompanied British imperialism strongly aided the metric cause in the long run. That nation's horrendous treatment of cultures in Asia, Africa, and elsewhere in the nineteenth century did much to destabilize indigenous cultures, disrupt habit and infrastructures, and wipe out local measuring systems, opening up the possibility of international consolidation around the metric system in the twentieth. Thus we return to the stories of the two very different measurement systems we saw earlier, in China and West Africa. One persisted for centuries because of its isolation, the other because of accommodation. The fate of each was similar, however; their experiences with nineteenth-century British colonization weakened them to the point where they could be replaced.

CHINA: OPIUM WARS

By the nineteenth century, British merchants had established trade routes to China, and Great Britain was looking for an excuse to demolish the highly restrictive trade regulations that the Qing dynasty imposed on foreign merchants. British vessels were only allowed in Canton—today known as Guangzhou—and British merchants were subject to Chinese law and not permitted to live in Canton itself. Meanwhile, British merchants had developed a thriving trade in opium, grown in India and exported to China. Horrified by this trade, which brought about the addiction of millions of China's inhabitants, the Chinese government tried to restrict it. In 1839, a new high commissioner for Canton, Lin Tse-hsu, tried to stop the opium trade altogether, and an ensuing series of incidents led to war. In one, the Chinese seized and destroyed 20,000 chests of British-owned opium; in another, Britain refused to hand over sailors who had murdered a Chinaman. Lin ordered all British expelled from Macao, next to Hong Kong; though Macao was a Portuguese trading port, China retained power in the region. In retaliation, Britain attacked and easily defeated Chinese soldiers in several port cities, including Canton, and began to march inland toward Nanjing before the Chinese capitulated. The terms of the ceasefire and subsequent treaty allowed the British to reside and trade in five ports, ceded Hong Kong to the British government, obligated the Chinese government to pay staggering sums of money (about half a billion dollars in today's terms) in "reparations" for the war it had caused by its crime of resisting British imperialism, and forced the Qing government to set up customs houses controlled by foreign governments. Two other wars followed in short succession, opening more cities to British merchants and those of other countries including France, Russia, Germany, and the United States, and establishing Macao as a free port, in what amounted to a foreign occupation of key parts of China. Its principal customs and markets were now in foreign hands, and the treaties forced foreign units on Chinese merchants in their dealings with other countries and diminished the authority of the emperor over weights and measures. Different countries stipulated different ratios of their weights and measures systems to those of China. These ratios, hated by the Chinese, were known as the Custom Ruler and the Custom Balance. The foreign weights and measures systems did not penetrate the countryside, but they did aggravate the already disorderly local measures. A Chinese scholar of Kula's ability and sensitivity would be required to describe the logic of variations of countryside units from region to region; that logic was now all but destroyed in the aftermath of the Opium Wars. As Guangming wrote in a book coauthored with other members of her group,

> The Qing government was neither able to resist the influx of these foreign systems and their applications in domestic daily life, nor did they have the power to unify the measuring and weighing systems in China. . . . The worst part was that the foreigners working in customs took the excuse that the Chinese measuring and weighing systems were too complex and chaotic, and there were no standards to observe; therefore, they had the right to establish regulations on their own, including the exchange rate.[16]

Foreign occupation had reduced the country to a semifeudal society and made the differences of its weights and measures a huge internal problem for China, in addition to all its other woes. Foreign customs officials distributed weights and measures standards of their own, forcing the Chinese to use them.

WEST AFRICA: THE ASHANTI WARS

By the nineteenth century, Great Britain had also come to exert a strong influence over the Akan region of West Africa, which it had named the Gold Coast. British trading settlements established in the region during the seventeenth century were now run by the African Company of Merchants. At first slaves were the principal source of wealth, but after the British abolished the slave trade in its colonies in 1807, attention turned to gold. The amount of "precious dust" that can be obtained by the "rude methods" of the natives has almost all been used up, says one article written in the nineteenth century, but "it is possible that European energy and skill might make it again a real gold coast."[17]

The area with the greatest potential for gold exploitation was controlled by a federation of Akan tribes called the Ashanti, whose capitol was at Kumasi. For a while, Great Britain was content to command a few forts in the Ashanti area, as did Dutch fellow colonizers, and remain aloof from the Ashanti, paying tribute to Ashanti tribes for the use of the forts. In 1821, this began to change. The British Government took over the settlements and forts from the African Company and seized more land to create the colonial territory of the British Gold Coast, whose seat was Cape Coast Castle. It stopped paying tribute to the Ashanti, which created friction with the tribes. In 1871, the British acquired the Dutch port of Elmina and refused to pay tribute for this as well. The Ashanti began to organize against the British, and the British sent a highly regarded army officer named Sir Garnet Wolseley to Cape Coast Castle.

In October 1873, at the beginning of what would become known as the Ashanti War of 1873–74, Wolseley defeated the Ashanti near

Elmina. In January, he set out inland for Kumasi, capturing it on February 4, 1874. The expedition was joined by several war correspondents, including a *Daily Telegraph* reporter named Frederick Boyle, who was attentive enough to note "scales and weights for gold dust" in abandoned towns the army marched through along the way, which he described as part of "the rubbishing treasures of barbarous life."[18] The British captured Kumasi on February 4. They promptly looted the city—taking not only gold ornaments and bags of gold dust but also the brass weights—blew up the royal palace, burned Kumasi to the ground, and returned to Cape Coast Castle with the loot, auctioning it at the end of the month. Boyle meticulously described the auction, even the brass weights: "They were cast in every possible form, fishes, and dragons, and gates, and swords, and guns, and insects, and animals. But the commonest was the human figure, male or female, in every possible attitude, in every operation of life. These brought £4. 15s to £3. 10s a dozen."[19]

Going through what they had plundered, the British discovered that cheating at weights and measures, found throughout Europe, was also practiced in Africa. Henry Brackenbury, Wolseley's military secretary, had the gall to reproach the Kumasi king whose city his army had annihilated:

> The gold-dust was carefully sifted by our valuer, and the dross removed from it; for it seems that adulteration, especially in gold-dust, is practiced to a very large extent in Ashanti. Indeed, amongst the loot we had brought away from the palace were several bags of fine brass-dust, which we had taken for gold-dust, but which afterwards turned out to be the spurious article used by the king to mix with real gold-dust when he makes payments.[20]

The Ashanti king, humiliated, was forced to sign a "Treaty of Peace" with the Queen of England, in which he agreed to pay a huge amount of "approved gold" to compensate England for the war the Ashanti had caused by the crime of opposing British aggression, and to allow free trade between the British and local merchants. Gold dust was outlawed as currency. In his book about Akan weights, Phillips writes, "The sack of Kumasi by British troops in 1874 sounded the death knell for the production and use of weights and the beginning of the end for gold dust as the monetary lingua franca of the Akan."[21] A century after the event, Phillips was thrilled to discover and purchase a copy of the *Daily Telegraph*'s coverage of the auction, mentioning the sale of the "quaint brass weights."

Kumasi's destruction broke the back of the Ashanti tribes, and little was done to repair it. In 1888, a British traveler was horrified to see the extent of the devastation wrought by his country's army on the legendary city. Everything had been destroyed, "the town was nothing more than a large clearing in the forest," and the inhabitants had lost all desire to rebuild. He describes:

> On all hands, amidst the universal ruin and decay, were hints of departed prosperity and evidences of a culture far superior to anything seen in the littoral regions; and as I looked round on the blighted city with its demolished buildings and its demoralized citizens, I could not help reflecting on the strange and regrettable fact that its ruin had been accomplished by a nation that yearly spends millions on the conversion of the heathen and the diffusion of civilisation.[22]

Twenty years after their first subjugation of the Ashanti capitol, the British were back to conquer it again. Claiming the Ashanti had not paid reparations, they marched into what remained of Kumasi and exiled the king. The use of gold weights was now banned, the British imposed the imperial system on the region, and another local measurement system, one of the most curious ever invented, was brought to a halt.

CHAPTER SEVEN

METROPHILIA AND METROPHOBIA

After the 1872 International Metric Commission, several prominent American scientists began to campaign for U.S. conversion to the metric system. Frederick Barnard (1809–1889), the president of Columbia College, was a leader of the movement. Barnard organized a meeting at Columbia College, which took place December 30, 1873, to form The American Metrological Society. He was elected president. Barnard also organized an educational version, the American Metric Bureau in Boston, which printed flyers and postcards to promote the system among the public:

> THE ENTIRE SYSTEM IN A SINGLE SENTENCE
>
> Measure all lengths in meters, all capacities in liters, all weights in grams, using decimal fractions only, and saying deci for tenth, centi for hundredth, milli for thousandth, deka for ten, hekto for hundred, kilo for thousand, and myriad for ten thousand.

Other helpful hints cited by metric reformers: the diameter of a nickel is 2 centimeters, and its weight 5 grams. Five nickels in a row equal a decimeter, two of them a decagram. Measures of capacity could be formed from the length measures. "Any person, therefore, who is fortunate enough to own a five cent nickel may carry in his pocket the entire metric system of weights and measures."[1]

During the next several decades, U.S. supporters of the metric system focused not on legislation but on public education, assuming that to be a necessary prerequisite to legislative action. There was pushback. Resistance began to grow among American engineers and many manufacturers, who realized that, however noble the sentiments, a U.S. metric conversion would impose a financial burden on them.[2]

In 1876, the Boston Society of Civil Engineers asked the Franklin Institute in Philadelphia to appoint a committee to hold hearings. The committee recommended against metric conversion. Practical life divided things into halves, quarters, thirds, and so forth, not decimals, the committee members pointed out, while all U.S. land surveys are in acres, feet, and inches. Over the decades, American industries had developed and acquired "an infinite variety of costly tools for working to exact measurements," which would make the cost of conversion enormously expensive. "If new weights and measures are to be adopted, all the scale beams in the country must be regraduated and readjusted; the thousands of tons of brass weights, the myriads of gallon, quart, and pint measures, and of bushels, half bushels, and peck measures, and every measuring rule and rod of every description throughout the land, must be thrown aside, and others, which the common mind cannot estimate, must be substituted." No doubt such a change seems easy to the "closet scholars who use weights and measures only in calculations," one engineer at the gathering remarked, "but to practical users of weights and measures, the producers and handlers of the material wealth of the country, the necessary cost of the change would vastly outweigh any possible theoretical benefit to be derived from it."[3]

•

In 1877, Congress issued a call for comments on a proposed bill for the United States to convert to the metric system. The responses were far less enthusiastic than the sponsors anticipated. Several Treasury Department officials were opposed, including even Carlile Patterson, who had replaced Benjamin Peirce as superintendent of the Coast Survey. Even more surprising was the fact that the negative report Patterson submitted had been drafted by none other than Julius Hilgard, the U.S. representative to the International Metric Commission. This report has been described by a pro-metric publication as "outstanding for its clarity, brevity, and eminent good sense," insofar as it earnestly and honestly assessed the impact of metrication on the Coast Survey, Treasury Department, and the general public.[4] In 1880, at the first annual meeting of the American Society of Mechanical Engineers, opposition to the metric system was a prominent theme.

Meanwhile, an extreme antimetric movement was born in Ohio and exhibited the classic signs of American antireform movements: xenophobia, rabid rhetoric, fabrication of "facts," reimagining history, conspiracy theories, and appeals to preserve the purity of nature and nation. The "enemy" was the "other": subversives, socialists, foreigners, atheists, and artifice. The good guys were patriots, capitalists, Christians, and adherents to God, country, and nature. Even back then, American antireformers tended to be colorful, eccentric people masquerading as populists, who traced their cause back to divine commandments and had wacky props.

METROLOGICAL MONUMENT: THE GREAT PYRAMID OF GIZA

The wacky prop of the American antimetric movement of the 1880s was the Great Pyramid of Egypt, the most bizarre example of an object seriously proposed as a metrological standard. The Pyramid stands in the middle of a desert, a few miles from the Nile River town of Giza. The oldest of the seven wonders of the ancient world, and the only one still intact, it has captivated the imaginations of visitors for thousands of years. Alexander the Great is said to have stood alone in the king's chamber after he conquered Egypt in the fourth century BC. During Napoleon's Egyptian campaign, which began in 1798, he too visited the Great Pyramid, and ordered his soldiers to wait outside while he entered the king's chamber alone. Scholars accompanying him issued voluminous reports on their return, firing the popular imagination for all things Egyptian, especially the Great Pyramid.

The pyramid's solidity and permanence are, in truth, the kinds of properties one looks for in a standard. John Herschel, a British amateur scientist who was on the commission to reconstruct the imperial standards, wrote, "Of human works, the most permanent, no doubt, and the most imposing . . . are those mighty monumental structures which have been erected as if for the purpose of defying the powers of elementary change."[5]

Early pyramid metrologists included Richard Vyse, a British Parliament member and army officer, who had been to Egypt, and his follower John Taylor, a partner in a London publishing firm, who had not. Both were convinced that secret knowledge was encoded in the Great Pyramid's dimensions, including the length of ancient measuring units. In 1859, Taylor published a pamphlet entitled *The Great Pyramid: Why Was It Built? And Who Built It?* followed in 1864 by *The Battle of the Standards: The Ancient, of Four Thousand Years, Against the Modern, of the Last Fifty Years—the Less Perfect of the Two.*"[6] Taylor decided that the pyramid's mathematical relationships were too sophisticated for ancient Egyptians to have grasped. The most obvious example was that the ratio of two sides of the pyramid's base to its height was exactly π, an irrational number unknown for centuries after the pyramid's construction. (Historians have suggested that since the Egyptians may have measured long horizontal distances by rolling a drum and counting the turns, they may have incorporated a relation to π into the structure while lacking specific mathematical knowledge of it.) Taylor believed, erroneously, that the Israelites had provided the slave labor for the pyramid construction (the Pyramid was built long before the Israelites were in Egypt). He asserted that its architectural design was the work of some Israelite (Noah was a leading candidate) who was following the instructions of the Great Architect (i.e., God) himself. Indeed, Taylor said, the pyramid had been built to provide "the measure of the Earth" for humans. Furthermore, the unit used to measure the stones was the "sacred cubit" (about 25 inches) that had been used by God's chosen people throughout the ancient world in their holy construction projects—Noah in his ark, Abraham in the tabernacle, and Solomon in the temple.[7] Furthermore, a coffer in the king's chamber was clearly a weight measure. The Great Pyramid, Taylor pronounced, was none other than the "alter to the Lord in the midst of the land of Egypt" of Isaiah 19:19 and provided the missing link between our measurement system and those of the Bible. The "battle of the standards" was therefore whether to use the ancient, sacred, and natural measurement system or the modern, artificial, metric one.

Taylor's pamphlets in turn inspired all kinds of pyramidologists, including a handful of professional scientists. The most prominent of these was Charles Piazzi Smyth, the Astronomer Royal of Scotland. Piazzi Smyth had participated in a meridian arc measurement and conducted astronomical observations at Tenerife in the Canary Islands. The arc of his career, however, took a dramatic turn for the worse after reading Taylor's *Battle* pamphlet in 1864, which convinced him of "the high probability" that "the Great Pyramid, besides its tombic use, might have been originally invented and designed to be appropriate for no less than a primitive Metrological Monument."[8] Piazzi Smyth then dashed off a book dedicated to Taylor, *Our Inheritance in the Great Pyramid*, which took the numerological claims up a notch. The fundamental pyramid unit, Piazzi Smyth declared, was not the cubit but its twenty-fifth part—the "Pyramid inch"—which was exactly 1/500,000,000 of the earth's axis of rotation. This unit, rather than the meter, was the true natural standard, pyramid measurements were "true cosmical relations in their original units," and the pyramid is "a Bible in stone, a monument of science and religion never to be divorced."[9] Anglo-Saxon peoples had wisely hewed closely to this measure throughout the years, for the present imperial inch only differed by a negligible fraction from the pyramid inch. Piazzi Smyth scorned the metric system, its inventors and its champions. "Simultaneously with the elevation of the metrical system in Paris, the French nation," he wrote, "did for themselves formally abolish Christianity, burn the Bible, declare God to be a non-existence, a mere invention of the priests, and institute a worship of humanity, or of themselves."[10]

Piazzi Smyth's book was a popular success when published in 1864, and he set off for Egypt to finally see the Great Pyramid for himself. There he found even more metrological wonders encoded in the pyramid's dimensions, including the number of days in a year, and the diameter and density of the earth. When he returned and presented these conclusions to the Royal Society, its members were unimpressed. They pointed out numerous errors in his work, including the fact that the famous ratio of twice one side to height was not equal to π after all, but to the more mundane ratio 22/7. The Egyptians may indeed have meant this to reflect the ratio of a circle's radius to circumference, but it did not signify knowledge of irrational numbers, nor divine guidance in the architecture. These and other errors made Piazzi Smyth's numerology collapse like a house of cards, but he continued to defend himself. As the ensuing controversy grew in intensity, so did Piazzi Smyth's confidence. He began making absurd comparisons: of himself to Kepler, and his opponents to the know-nothings who had ridiculed Kepler. In a fit of rage following a tangle with none other than James Clerk Maxwell, Piazzi Smyth resigned from the Royal Society in 1874.

Yet a few years later, Piazzi Smyth found followers in the United States, among the antimetric movement of the "International Institute for Preserving and Perfecting the Anglo-Saxon Weights and Measures" in Ohio, with whom he corresponded. At last he had sympathetic ears. The first was Charles Latimer (1827–1888), a chief engineer for Atlantic and Great Western Railway whose interests veered toward the occult. In the late 1870s, Latimer discovered the works of Piazzi Smyth, whose vision was indeed bizarre—God had seen that His fundamental unit was incorporated by a Hebrew architect into an Egyptian pyramid in order that the English people could incorporate it into the imperial system—and was excited by them. Piazzi Smyth's numerological claims about the pyramid had been thoroughly discredited by the late 1870s, but his books about the pyramid continued to sell well and inspire mystics and numerologists such as Latimer. The Great Pyramid of Egypt became an American emblem for the Ohio organization, whose members interpreted the pyramid as being the same as the one on the Great Seal of the United States, which appears on the reverse of every dollar bill.

In 1879, Latimer composed a pamphlet called *The French Metric System; or the Battle of the Standards*,[11] whose cover was decorated with the great seal. The battle that *Nature* had declared over was not; it was just beginning, Latimer declared. Don't give up now! A worldwide conspiracy of atheists is against us—and the symbol of our resistance is the pyramid. The pyramid, he wrote, supplies "the true solution of the questions agitating the world upon weights and measures." In it we find that the inch—and the cubit containing twenty-five of them—is natural, commensurate with the earth, and of divine origin. By contrast, the metric system is unnatural, incorrect about the earth's dimensions, and invented by atheists. As for the fact that more and more countries were adopting the metric system, Latimer quoted a correspondent: "If other countries are going helter-skelter down the road to atheistical ruin, it is happily the spirit of the Pilgrim Fathers which keeps, and we may hope always will keep, America the last in such a negative and suicidal race as that."[12]

On November 8, 1879, at high noon in the Old South Church of Boston, Latimer and two compatriots launched an organization whose full name was the "International Institute for Preserving and Perfecting the Anglo-Saxon Weights and Measures, and for Opposing the Introduction of the French Metrical System amongst English-speaking peoples."[13] The choice of Boston, and the Old South Church, was calculated to underscore their patriotic zeal and to help ensure international connections. Latimer was actually from Cleveland, Ohio, home of the organization's major branch. The society met every two weeks in Cleveland, and held an annual meeting in November.

The movement's literary organ was the *International Standard*, which published its first issue in March 1883 and was described on its cover as "A Magazine Devoted to the Preservation & Perfection of the Anglo-Saxon Weights & Measures and the Discussion and Dissemination of the Wisdom Contained in the Great Pyramid of Jeezeh [Giza] in Egypt." A global conspiracy, Latimer declared in the introduction to the first issue, was trying to force the metric system, an evil "born of infidelity and atheism," on the Anglo-Saxon world, and all patriots must organize to fight back. Contributors to the *Standard* made much of the fact that the scientists had blundered in trying to base the meter on a natural standard, while the pyramid's units were divine. They also denounced the tyranny of the government forcing something unwanted on Americans. Subsequent issues contained numerological studies of the Great Pyramid, rants against the metric system, supportive letters, refutations—denunciations, actually—of opponents, poems, and even anti-metric songs.

Charles Totten (1851–1908), an artillery lieutenant, was a frequent contributor who dedicated his book *An Important Question in Metrology, Based Upon Recent and Original Discoveries* to members of the Anglo-Saxon race.[14] We Anglo-Saxons, he said, can trace our ancestry to the two adopted sons of Joseph, specially designated by God to be a "great people." But to maintain our greatness, we have to defend our "native Anglo-Saxon metrology," which derives from the "God-designed metrology of Israel" found in the Great Pyramid and which needs but slight rectification (i.e., to be brought back into line with the pyramid inch) to make it "absolutely perfect." Totten, too, was deeply into numerology and, by superimposing the pyramid's proportions over a figure of

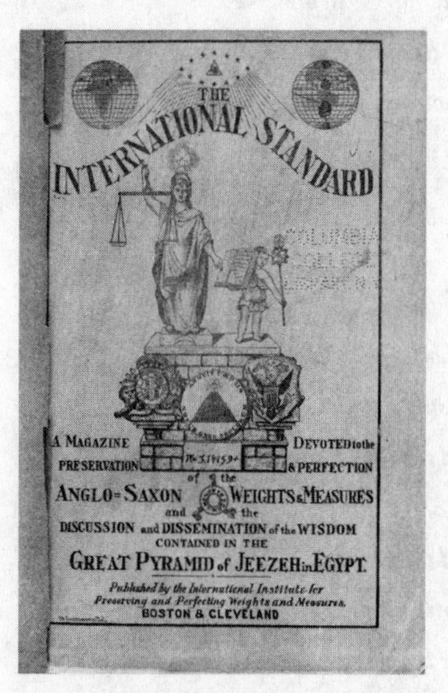

COVER OF THE FIRST ISSUE OF THE *International Standard*.

the human body, he demonstrated that its passageways correspond to the locations of "the womb, the heart, and the lungs." Trying to change the Anglo-Saxon system of measurement, Totten declared, would be unconstitutional; Article I, Section 8, merely gives Congress the right to fix the standards, not to choose the units, which is "the people's question."

Totten's more unusual contributions to the *International Standard* include a song, "A Pint's a Pound the World Around (Remove not the ancient landmarks)," that was published complete with music, with lyrics he had composed:

A pint's a pound the world around,
We Anglo-Saxons claim,
So long it's stood, to make it good
We Anglo-Saxons aim.

The "ancient landmarks" to preserve
We've firmly set our face;
They show the footprints o'er the earth,
Of Khumry's wand'ring race.

(Chorus)

Then swell the chorus heartily,
Let ev'ry Saxon sing,
"A pint's a pound the world around,"
Till all the earth shall ring;
"A pint's a pound the world around"
For rich and poor the same:
Just measure and a perfect weight
Call'd by their ancient name.

Subsequent choruses celebrate the "old traditions," denounce the "metric scheme," and appeal to "our Father's 'rule.'"[15]

Pro-metric advocates, however, failed to make headway in the 1880s, and no legislation promoting the metric system was passed. This leached much of the energy from the movement, which began to focus more on pyramidology. Latimer's death in 1888 removed what little energy remained, and the *International Standard* ceased publication. When Piazzi Smyth died, in 1900, a stone pyramid and a cross marked his grave.

The history of metrology includes efforts to tether units to artifacts, natural phenomena, and physical constants, but this is one of the few attempts to carry out metrology by divine revelation. Strangely enough, metrology by revelation may be the most uncertain of all. As the American philosopher Charles S. Peirce warned, apropos of attempts to use revelation to gain knowledge, "We do not know His inscrutable purposes, nor can we comprehend His plans. We cannot tell but He might see fit to inspire His servants with errors."[16] The Great Pyramid episode is fascinating in the way it exhibits in an extreme form the passions that tend to crystallize around the quest for metrological finality.

METRICITIES AND METRIC FALLACIES

The first director of the National Bureau of Standards (NBS) in the United States, Samuel W. Stratton (1861–1931) was an aggressive, far-sighted scientist who remained in that post for a remarkable 22 years, from 1901 to 1923. In 1905, he held a meeting of state weights and measures officials that uncovered "the greatest diversity" in state weights and measures; the NBS therefore resolved to seek "uniform laws and practices" relating to these matters. This grew into an annual conference that continues today, the National Conference on Weights and Measures.

In 1902, Stratton was instrumental in getting a metric bill introduced into Congress. The Mendenhall order, the creation of the NBS, and new American political dominance over Cuba, the Philippines, and Puerto Rico—all of which used the metric system—seemed to make metric conversion more compelling. Hearings continued into 1902. Lord Kelvin himself—science royalty—crossed the Atlantic to testify in its support and said that if the United States were to adopt the system, England would surely follow.

The pro-metric attention provoked a backlash again, this time led by two members of the American Society of Mechanical Engineers, Samuel S. Dale and Frederick A. Halsey. The two were comically mismatched—Dale was voluble while Halsey's behavior tended to be scholarly—and they quarreled about virtually everything, but their adherence to this cause prompted them to swallow their personal animosity. In 1904, they jointly published a book that consists of two parts, one by each author. The title page reads: "The Metric Fallacy, by Frederick A. Halsey, and The Metric Failure in the Textile Industry, by Samuel S. Dale."[17] In his introduction, Halsey contrasts the Anglo-Saxon nations, who are the "only ones that have ever dealt with the subject of weights and measures in a rational manner," and who have constructed the "simplest and the most uniform system of weights and measures of any country in the world," with the French, whose idealistic, "rainbow-chasing" system has never "made material progress in industry except when backed by the policeman's club." Above all, he feared the impact on the English-speaking peoples, who "have built up the greatest commercial and industrial structure the world has known." Halsey continues, "They are asked to destroy the very warp and woof of their own vast industrial fabric in order that they may assist in weaving another of alien origin and with no resulting gain except to aliens."[18]

The book, evidently funded by the American Society of Mechanical Engineers, was immensely popular, "the most influential single anti-metric work ever published."[19] Its success was due in part to the doggedness of the authors. Dale in particular gave new meaning to the expression "indefatigable." No one was too important or unimportant for him to hound; he sent letters to schoolteachers and small-town newspaper editors as readily as he sent letters to senators, cabinet secretaries, and presidents, to protest any perceived pro-metric sentiment. Dale wrote President Theodore Roosevelt protesting the allegedly sloppy treatment of the nation's yard standard on exhibit at the St. Louis Exposition in 1904 (it turned out to be a copy), using the letter as an entrée into a denunciation of what he called "the Mendenhall conspiracy"—he called it illegal and unconstitutional—and any attempt to metricize the United States.[20] Dale obsessively collected books about weights and measures, and every scrap of correspondence, and gave his collection of 1800 books dating back to 1520 and tens of thousands of pages of correspondence to Columbia University.[21] The formidable opposition to which Dale and Halsey contributed prevailed, the metric advocates gave up, and the pair went their separate ways.

World War I fostered a spirit of international cooperation and solidarity in the Americas that went under the name of Pan-Americanism. This reinvigorated metric advocates—by this time South American countries were solidly metric—who had another metric bill, called the Ashbrook Bill, introduced into Congress in 1916. In 1916, a pro-metric group called the American Metric Association formed, with Stratton on its executive committee. Dale and Halsey overcame their differences to reunite in founding an antimetric organization, the American Institute of Weights and Measures, again with the backing of the American Society of Mechanical Engineers. They still quarreled vociferously and continuously (Dale carefully preserved their melodramatic exchanges accusing each other in detail of betrayal) but the "cause" kept them doggedly working together in their battle against those whom they derisively referred to as "metricities." Dale and Halsey denounced the Ashbrook Bill as unconstitutional—fixing weights and measures belongs to the states—and as damaging to American business and engineering. Dale wrote a pamphlet called *The Mendenhall Conspiracy to Discredit English Weights and Measures*, which charged that the treasury secretary's 1893 act was illegal and unconstitutional; only Congress has the right to fix standards. Insults energized them. When the American Metric Association President George Kunz wrote to Dale thanking him for sending a pile of antimetric publications and for revealing what it was like to be a "moron," Dale frequently quoted the remark to others, commenting that the insult clearly demonstrated that Kunz was a "boor and a bigot."[22]

This time, patriotism and pocketbooks, not pyramids, dominated the rhetoric. "It is written in the stars that in [the] future this is to be an Anglo-American world," Halsey wrote to one correspondent. "Let us make it Anglo-American in its weights and measures."[23] Each time a bill was introduced into Congress, the Institute would rouse its members against it. "The manufacturers of the country, whose pocketbooks will be threatened, whose business interests will be jeopardized, must become active in their own defence [*sic*]," says one Institute letter summoning its members to pack Congressional hearings.[24]

Dale and Halsey struggled to keep religion, yet another of their differences, out of the conversation. They were not always successful, particularly with respect to evolution and creationism. In 1925, Halsey fired off a rebuttal to a mildly pro-metric editorial in *The New York Times*. The letter was published, as it happened, a few days after the end of the Scopes trial. Entitled "Metric System a Failure," Halsey's letter could not resist a jab at antievolutionists, whom he viewed (like the metricities) as blinded by their ideology, and ended his letter as follows: "The case for the English system of weights and measures is as thoroughly established as that for evolution, but the metric party is as blind to proof as are the Fundamentalists. Indeed, the metric advocates are the Fundamentalists of Science. Obsessed with a hobby, they do not know and they will not learn."[25]

Dale, who did not believe in evolution and thought its advocates were blinded by ideology, was outraged. He wrote to Halsey, chastising him for bringing up this dangerous subject that they had agreed to keep off the table. He said that if he must refer to the "fundamentalism versus evolution" controversy, the truth is that metric system propaganda is distributed by "so-called scientists, who have defied the elementary principles underlying true science by suppressing and ignoring plain facts and deliberately insisting on propositions plainly erroneous," whereas real scientists "can be actuated by the love of truth and the readiness to accept truth regardless of consequences which are essential qualifications in the determination of the truth regarding the origin of the varied forms of life on this planet."[26]

Congressional hearings on the metric system were held in 1921 and 1926, but again no action was taken. In the 1930s, too, little legislative action was attempted. Sports was one of the few issues keeping the metric system in the public eye; as international competitions grew and more countries turned metric, U.S. athletes found themselves in competitions where the long jumps, high jumps, track and pool lengths were in metric units, sparking a controversy among U.S. amateur sports officials as to whether they should not convert to the metric system ahead of everybody else. At the 1936 Olympics in Berlin, following the gold medal–winning long jump victory by African American athlete Jesse Owens, a *New York Times* reporter who liked to skewer sports antimetricists asked which was the more impressive way of stating Owens's record-breaking achievement: 8.06 meters—or 8 yards, 2 feet, 5 and 21/64 inches?[27]

VESTIGIAL MEASURES

For a quarter century, Dale and Halsey repeatedly and vociferously hammered away at their claim that the metric system had failed and that metrical confusion was rampant throughout Europe, its supposed home.[28] They relied mainly on the testimony of sympathetic correspondents rather than independent research. In 1926, Arthur Kennelly (1861–1939), a professor of electrical engineering at Harvard University and an executive committee member of the American Metric Association—and one of hundreds of recipients of Dale diatribes—was up for a sabbatical. He decided to spend it in Europe to see the situation for himself. In the past quarter century, he said, the voluntary self-imposition of the metric system in over thirty countries, with a combined total population of over 300 million people, provides us with the opportunity for a grand sociological experiment. "Some allege," he wrote, footnoting Dale and Halsey, "that the metric system has failed in Europe, and that the old units are the ones generally used." He published his findings in *Vestiges of Pre-metric Weights and Measures Persisting in Metric-System Europe, 1926–1927*.[29]

Kennelly indeed found vestiges of traditional measurement units, though they were rare and their use was ebbing. He found archaeological artifacts, of course. Old church and municipal buildings, for instance, might have ancient length standards cut into their sides for communal use, or road markers might be labeled with old distance units, "accidental residuals from a long forgotten past." In the Roman Forum, he found the Milliarum Aureum, or "golden milestone" that Emperor Augustus had set up as the zero or starting point of all roads throughout the empire.

Kennelly found some seemingly traditional units in use that were in actuality metric units or their subdivisions with the old names attached; these he called metricized or submetricized units. In many French-speaking lands, for instance, people referred to the *livre*, meaning half a kilogram (though 400 grams in Marseilles); in German-speaking lands, the same was true of the *Pfund*. Others included the *lieue* (which now meant 4 kilometers); the *toise* (2 meters); the *once*, or *ons* (25 grams); and the *pied de neige*, or foot of snow (a third of a meter of snow). But the persistence of these traditional names did not mean that the old units were still current, wrote one of Kennelly's correspondents, any more "than that the Hotel Powhatan at Washington, DC, is kept by aboriginal Indians, because of its Indian name."[30]

Kennelly found some of the old units peacefully existing with the new. Hats, shirts, and boots might be labeled with two sets of numbers, designating their sizes in the metric system and the imperial system. Kennelly decided that these do not really count as "infractions of the metric system" because often neither buyer nor seller knows the significance of the numbers, but regards them as trade sizes rather than measurements. Other nonmetric units were tolerated because they were used by the elderly. One French scientist told Kennelly of a time he caught a severe chest cold in Menton, France, and visited a physician who happened to be Scotch. The physician wrote a prescription in the imperial measure, which the French pharmacy followed—accurately, the scientist made sure. Back in Paris, the scientist complained about this violation of French law, but was told that these infractions were mainly harmless and difficult to enforce.

Kennelly also found improvised units used where no metric unit seemed appropriate. These included the *soma*, used in the Italian region of Umbria, to designate the quantity of firewood that a mule or donkey can carry on its back, and the Swiss *Stunden* or hours used in road markers—travelers in that mountainous country found it much more sensible to mark out distances for walkers in hours rather than linear units, given the difference in speed when one was going uphill or down. Certain German towns still used the *Lot*, referring to a measure of coffee beans about the size of a cup, while in Bologna the old *libbra* and *oncia* were still sometimes used to measure silkworm eggs, and the *castellated* and *carro* to measure grapes and wood. In Mallorca, silver was weighted and sold by the traditional *adarme*. But these traditional uses, too, he thought harmless.

Kennelly discovered that many traditional units had recently vanished, but not because of the metric system. In France, the once-popular league or *lieue*, which expressed the distance of one hour's walk, had vanished because of the bicycle's popularity. The *corde*, an amount of firewood, persisted in areas that relied on wooden stoves but was disappearing in areas with central heating. Metrication, in short, benefited as much from changes in the European environment as from official acts.[31]

But Kennelly thought that the most important cause of the decline of traditional measures—apart from laws mandating use of the metric system—was the growing interdependence of these regions. "I can imagine that in the middle ages, these little villages regarded themselves as relatively remote," wrote one of his French correspondents, "just as we in our own time look upon ourselves as remote from America. However, the recent exploit of Lindbergh [who had crossed the Atlantic solo just three months previous to this letter] should make us reflect. We must hope for a time when all peoples will adopt one and the same system."[32]

Dale and Halsey were wrong, Kennelly concluded. The metric system had rooted firmly in Europe, and was growing steadily stronger. When traditional units still appeared, it was rare and due to archaeology, sentiment, or local improvisation.

CHAPTER EIGHT

SURELY YOU'RE JOKING, MR. DUCHAMP!

It's a "joke about the meter."

—M. DUCHAMP

In the galleries on the fifth floor of the Museum of Modern Art (MoMA) in New York City, you can see some of the most famous artwork of the late nineteenth and early twentieth century. Vincent van Gogh's *Starry Night* hangs in one room, Salvador Dalí's *The Persistence of Memory* in another, Henri Matisse's *Dance* in a third. You walk through a roomful of Piet Mondrian compositions, including *Broadway Boogie Woogie*, and it seems like every other painting is by Pablo Picasso, including *Les Demoiselles d'Avignon*.

Off in one corner, at the top of a staircase, a small gallery contains works that were composed quite differently. In the room's center is *Bicycle Wheel*, consisting simply of a bicycle wheel mounted by its fork upside down on a kitchen stool. It was created by the French artist Marcel Duchamp (1887–1968), who is most famous for controversial works such as *Fountain*—a porcelain urinal that Duchamp signed with an assumed name—and *L.H.O.O.Q.*, a Mona Lisa reproduction on which

MARCEL DUCHAMP.

he drew a moustache and goatee beard. *Bicycle Wheel* is what Duchamp called a "ready-made," an ordinary object that he had selected, assigned a title, and declared to be art.

Along one wall, in a glass case, is another intriguing Duchamp work called *3 stoppages étalon* (or, in English, *3 Standard Stoppages*). It consists of an old wooden croquet box with its top open. Inside are not mallets but what look like two long and thin glass "slides," each with a "specimen" consisting of a long, snaking piece of thread mounted on canvas. The box also contains two broken wooden slats. Above it hangs a third thread-bearing slide and a third broken slat. The museum's label next to the case dates the piece as constructed between 1913 and 1914, and states the following: "It is a 'joke about the metre,' Duchamp glibly noted about this piece, but his premise for it reads like a theorem: 'If a straight horizontal thread one metre long falls from a height of one metre onto a horizontal plane twisting *as it pleases* [it] creates a new image of the unit of length.' "

The description continues: "Duchamp dropped three threads one meter long from the height of one meter onto three stretched canvases. The threads were then adhered to the canvases to preserve the random curves they assumed upon landing. The canvases were cut along the threads' profiles, creating a template of their curves—creating new units of measure that retain the length of the meter but undermine its rational basis."

The Stoppages, as it is nicknamed, is a strange piece and the accompanying label makes an alarming claim. Does this artwork really sabotage the rationality of the meter—that fundamental SI unit which its creators, the French revolutionaries, viewed as scientifically and politically liberating? Or is it another Duchampian gag, the irony of which the curators missed, in which the artist mocks the obsession with precision and universalism which he was sensitive enough to notice in the culture in which he grew up? The casual visitor might even wonder what science is doing in the artist's work at all. What did Duchamp know about science, the metric system, and metrology, and what motivated him to produce this object? For an answer, we have to return to the early years of the twentieth century, a time when the scientific world was in turmoil and Duchamp was growing up.

SCIENCE ANGST

At the beginning of the twentieth century, startling scientific discoveries (X-rays, radioactivity, the electron) and powerful new technologies (electrification, wireless telegraphy) were radically transforming human life and our perception of nature. Although he was a painter by training—he followed his two elder brothers by moving to Paris at the age of seventeen to become an artist—Duchamp and other nonscientists were able to keep abreast of the scientific world thanks to many high-quality science popularizations of the day. Scientists like Marie Curie

DUCHAMP'S *3 stoppages étalon* (*3 Standard Stoppages*).

and Ernest Rutherford wrote summaries of their research in popular magazines, while others, such as Jean Perrin and Henri Poincaré, wrote best sellers. Science fascinated popular culture.

Poincaré's books included his 1902 effort *Science and Hypothesis*, which went through twenty editions by 1912. He wrote eloquently of how recent developments were shaking the foundations of Newtonian mechanics and generating doubts about the very notion of scientific objectivity. Poincaré advocated a philosophical position known as "conventionalism," which conceived of geometries, and indeed all scientific laws, as mere conveniences—mental projections or frameworks—rather than actual descriptions of nature, an idea that would profoundly influence artists such as Duchamp.

The ideas conveyed by such books provoked a cultural anxiety. On the one hand, science seemed to promise stability—an orderly and universal mechanical picture of the world, reliable and powerful technologies, and increasing material comforts—and to restructure global society, thanks to an expanding international collaborative network of institutions and treaties. Yet as Europe slowly slid toward the Great War, there were many indications that this promise might be unfulfilled; the mechanical picture was not as free of chaos, the technologies not as innocent, and the comforts not as secure as they seemed.

The French poet and cultural critic Paul Valéry (1871–1945) expressed this ambivalence in paradoxical language. Our passion for order, he wrote, was creating chaos, our virtues were promoting horrors, our rationality was fostering irrationality. Valéry did not specify an example but may well have been referring, for instance, to the concept of "the quantum," which emerged from the drive to tidy up the fine details of thermodynamics. The poet wrote that the "rage for precision" was leading to the opposite, to a state in which "the universe is breaking up, losing all hope of a single design, to the ultramicroscopic being much different from the everyday world, and to determinism causing a crisis in causality."

Valéry's claim that "unpredictability in every field is the result of the conquest of the whole of the present world by scientific power" was even bolder. In his eyes, the impact of science was being felt in all areas of human life, and he predicted "an amazing change in our very notion of art."[1] Many artists of the time indeed found acquaintance with science indispensable: its provocative discoveries were front-page news and challenged deeply held notions of reality.

One hot topic was the fourth dimension, conceived as another spatial dimension rather than as time (as it would be after the 1919 confirmation of general relativity predictions). Many novelists, musicians, and painters found this idea stimulating, even liberating. For some, it suggested a new space in which the world can be seen, for others the existence of a multiplicity of perspectives, and for yet others the existence of orders of reality that artists alone could intuit and reveal. Much of the influence of the fourth dimension was due to the discovery of X-rays, which made the existence of invisible structures of reality just outside our vision no longer a philosopher's metaphysical concern or an occultist's fantasy, but scientific fact.

METER MADE

Having exhibited an early aptitude for drawing and mathematics, Duchamp's scientific interests were already apparent even in his earliest works. For example, his *Nude Descending a Staircase #2* (1912) represented a moving body by abstractly decomposing the form into a succession of plane surfaces. But Duchamp was shocked when his more well-connected siblings asked him to withdraw that work from an independent art exhibition they were to be in together. "[A]s a reaction against such behaviour coming from artists whom I had believed to be free," he told one interviewer, "I got a job. I became a librarian at the Sainte-Geneviève Library in Paris."[2]

While he worked in the library from late 1912 to 1915, Duchamp read extensively about art and popular science. He also visited the Musée des Arts et Métiers, a popular destination for those interested in science and technology, which has a wonderful collection of weights and measures. "I had no intention of having shows, or creating an *oeuvre*, or living a painter's life,"[3] he recalled much later. Duchamp ruminated about how to make art meaningful in a science-rich cultural environment. His voluminous notes, later published, reveal careful readings of scientific literature on the fourth dimension, non-Euclidean geometry, electricity, phase transitions, thermodynamics, radioactivity, atomic structure, biology, and more.

"Make a painting *of frequency*," reads one note, while another refers to a "painting of precision," and another to a quest for "playful physics." Yet another note talks about the desire to create "*a reality which would be possible by slightly distending* the laws of physics and chemistry."[4] These—and other—notes show Duchamp's fascination with the clear, spare prose of science, with the experimental method, with chance, with the bivalent character of precision mentioned by Valéry, and with Poincaré's conventionalist philosophy.

Conventionalist philosophy helped Duchamp break free of traditional aesthetics—or what he would dismissively refer to as "retinal art," "taste," and "arty handiwork." He regarded its principles not as fundamentals but as choices, and sought alternatives. After a visit to the Salon de la Locomotion Aérienne at the Grand Palais in 1912, in the company of fellow artists Fernand Léger and Constantin Brancusi, Duchamp remarked, "Painting's washed up. Who'll do any better than that propeller?"[5] Duchamp moved to an apartment near the library where he conducted experiments in putting lines on canvas in different "dry" ways, such as by mechanical means and chance effects.

These concerns would culminate in 1923 in what art historians regard as Duchamp's masterpiece, *The Bride Stripped Bare by Her Bachelors, Even* (commonly known as The Large Glass). *The Bride Stripped Bare*, consisting of two huge framed panes of glass (shattered in an accident but repaired by Duchamp), standing vertically, between which are sandwiched various mechanical objects, geometrical forms, and pieces of wire, is a brilliant and iconoclastic work that owes much to Duchamp's meditations about science. En route to this intellectual-artistic project, Duchamp performed experiments and composed works that employed chance, the fourth dimension, and elements of science playfully wielded in a way that provocatively sought to blur the difference between art and everyday objects.

Erratum Musical (1913) was a vocal piece in which Duchamp's sisters sung notes pulled randomly from a hat. Duchamp's ready-mades (or "tout fait" in French) were ordinary objects that had been taken from their functional context, assigned a name, and placed alongside other art objects to become artwork themselves. Other works poked fun at concealed codes and networks that structure the world, making them visible in his works and revealing them to be not permanent fixtures but only conventions. For example, *Tzank Check* (1919) is a playful composition used to pay his New York dentist Daniel Tzank, a drawing of a $115 check drawn on the fictional "Teeth's Loan & Trust Company." *Monte Carlo Bond* (1924) is a drawing based on a traditional bond design, which Duchamp copied and sold as part of a mathematical scheme to win at Monte Carlo's roulette tables. These two simultaneously parodied and called attention to how pieces of paper, given the proper conventions, became an exchange for services.

Duchamp's impulses—the playful approach to science, the conventionalist stance, the parodies of the earnest precision of others, and the irreverence toward societal agreements that structure the world—intersect in the Stoppages.

Duchamp was not the first artist or author to spoof standards. The writer Alfred Jarry (1873–1907), a progenitor of surrealism and Dada, and a prankster like Duchamp, often referred to contemporary physicists, including C. V. Boys, William Crookes, Lord Kelvin, and James Clerk Maxwell, in his works. In 1893, Jarry coined the term *'pataphysics* (the initial apostrophe is intentional) to describe his own playful treatment of science, which included the spoofing of the obsessive preoccupation with measurement standards. For instance, Doctor Faustroll, the protagonist of Jarry's novel *Exploits and Opinions of Doctor Faustroll, 'Pataphysician* (published 1911), carries in his coat pocket a centimeter that is "an authentic copy in brass of the traditional standard."

To be sure, it was an age of creative spoofing. *Merle Blanc*, a French humor magazine, conned the curator of the Versailles Palace into extending an invitation to examine a rare "double decimeter measure in rosewood" that had been owned by Madame de Pompadour (Pompadour had died 30 years before the metric system was invented and thus before decimeters existed). The magazine then set about trying to convince French museums to accept donations of Napoleon's automobile, a bracelet owned by the Venus de Milo, and Victory of Samothrace's eyeglasses.[6]

Duchamp's spoof of standards was richer, wittier, more deeply engaged with science than these, and centrally connected with his own artistic concerns. He began to create what would become the Stoppages around 1913–14. In one of Duchamp's notes, he writes that "if a straight horizontal thread one metre long falls from a height of one metre onto a horizontal plane distorting itself *as it pleases*," it creates "a new shape of the measure of length." Another note called it "canned chance." In 1914, again according to the notes, Duchamp repeated this experiment three times, fastening the threads to pieces of canvas.

Duchamp took the three canvases with him when he left Paris for New York in 1915. In 1918, while working on his last oil painting, *Tu M'*, he had wooden templates cut to the shape of the curves, creating parodies of "metre sticks" and objectifying the standardization of measurement in graphic form. In 1936, further revising what he had done two decades previously, he affixed the canvases bearing the threads to three glass plates and put them in a wooden box along with the three wooden rulers. The work was first shown at MoMA in an exhibition in 1936–37, and its photo first appeared in an art journal in 1937.

A *stoppage* is something stopped or brought to a halt; *3 stoppages étalon* is three stopped thread-standards. Over the years, Duchamp told interviewers that the work was his first use of chance, a step away from artistic technique, and a "forgetting of the hand." It was an important effort, he said he had come to realize, "liberating me from the past." In response to a questionnaire that Duchamp filled out when the work entered MoMA's collection in 1953, he wrote, "A joke about the metre—a humorous application of Riemann's post-Euclidean geometry which was devoid of straight lines." Art historians have studied the role this work played in Duchamp's later compositions, and in particular on his separation from traditional art. Art historian Francis Naumann contributed a pioneering study of the Stoppages in *The Mary and William Sisler Collection*, published by MoMA in 1984, and pointed to the influence on Duchamp of the philosopher Max Stirner. In the next decade, University of Texas art historian Linda Henderson explored the connection between the Stoppages and Duchamp's interests in non-Euclidean geometry and contemporary metrology. "In subverting the standard meter by generating three new inconsistent standards of measure," Henderson wrote in her 1998 book *Duchamp in Context*, "Duchamp had moved beyond traditional definitions of art itself."[7]

About a decade ago, the story of Duchamp's *3 stoppages étalon* took a bizarre twist. New York City artist Rhonda Shearer and her late husband Stephen Jay Gould—the Harvard paleontologist, evolutionary biologist, and author—were tracking down the exact sources of Duchamp's ready-mades and other works. They attempted to replicate *3 stoppages étalon*, and dropped meter-long threads to the ground; they found that their threads twisted and kinked, landing in anything but smooth curves. Examining the work at MoMA, they noticed that the threads are not *exactly* a meter long and are stuck through pinholes to the other side of the canvas, where they continue for a few centimeters and are sewn in place.

Shearer and Gould concluded that, thanks to their experiments, they had uncovered Duchamp's real motive. This, they claimed, was to deceive people about how he manufactured the work, knowing the deception eventually would be exposed by those who used the scientific method, teaching a lesson about perception. They created an online journal devoted to Duchamp (toutfait.com) and founded an Art Science Research Laboratory to explore the use of "scientific method in the humanities" and promote the idea that "methods, not people, are objective."[8]

Art historians were unimpressed. They found it unsurprising that Duchamp, who was irreverent about standards to begin with, loved a joke, and had spoken of his desire to "distend" the laws of physics, had used threads of uneven length. Nor were they perturbed that the threads passed through holes and were sewn to the canvas on the other side; Duchamp used a similar method to fix lines to canvases in other works, and may have planned to fix the threads this way from the start (for which purpose, furthermore, he would have had to have used threads a little longer than a meter). The art historians found the philosophical issue behind the Stoppages more relevant to understanding it than the length of its threads.

As for the question of the way Duchamp had gotten the threads to curve smoothly, Jim McManus, an emeritus professor of art history from California State University, Chico, asked an elderly German tailor who had been trained in traditional European methods about threads and techniques. The tailor mentioned a common practice of using buttonhole twist thread, which was waxed to give it additional strength, as the foundation for suits and garments. McManus therefore purchased some threads and wax, experimented with Naumann, and found that they were able to produce the kind of results Duchamp obtained. They concluded that it was likely that Duchamp had indeed created the work in the way he'd said.

McManus then concocted a spoof of his own. He created a "Do-It-Yourself Home *3 Standard Stoppages* Starter Kit," containing thread, wax, and some instructions. He issued it under the pseudonym "Rrose Sélavy," which Duchamp used as his female alter ego (a pun of its own, sounding like "*Eros,* c'est la vie"). The kit promised to permit its pur-

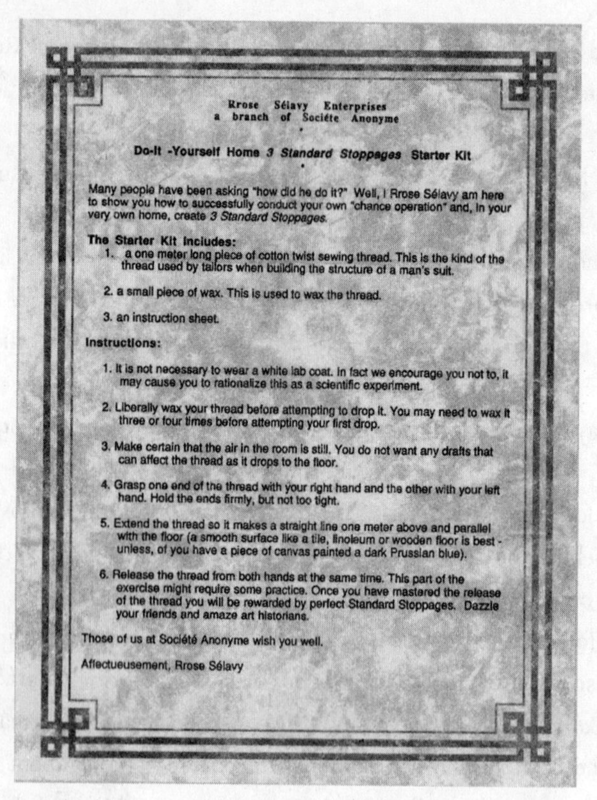

Rrose Sélavy Enterprises
a branch of Sociéte Anonyme

Do-It -Yourself Home *3 Standard Stoppages* Starter Kit

Many people have been asking "how did he do it?" Well, I Rrose Sélavy am here to show you how to successfully conduct your own "chance operation" and, in your very own home, create *3 Standard Stoppages.*

The Starter Kit Includes:

1. a one meter long piece of cotton twist sewing thread. This is the kind of the thread used by tailors when building the structure of a man's suit.
2. a small piece of wax. This is used to wax the thread.
3. an instruction sheet.

Instructions:

1. It is not necessary to wear a white lab coat. In fact we encourage you not to, it may cause you to rationalize this as a scientific experiment.
2. Liberally wax your thread before attempting to drop it. You may need to wax it three or four times before attempting your first drop.
3. Make certain that the air in the room is still. You do not want any drafts that can affect the thread as it drops to the floor.
4. Grasp one end of the thread with your right hand and the other with your left hand. Hold the ends firmly, but not too tight.
5. Extend the thread so it makes a straight line one meter above and parallel with the floor (a smooth surface like a tile, linoleum or wooden floor is best - unless, of you have a piece of canvas painted a dark Prussian blue).
6. Release the thread from both hands at the same time. This part of the exercise might require some practice. Once you have mastered the release of the thread you will be rewarded by perfect Standard Stoppages. Dazzle your friends and amaze art historians.

Those of us at Société Anonyme wish you well.

Affectueusement, Rrose Sélavy

ART HISTORIAN JIM MCMANUS'S
"DO-IT-YOURSELF HOME *3 Standard Stoppages* STARTER KIT."

chaser to "Dazzle your friends and amaze art historians" by creating *3 Standard Stoppages* "in your very own home."

McManus once gave me a starter kit, and one day my son and I tried it out. First we tried dropping a meter-long length of thread without waxing it; the thread fluttered down, looped and twisted, and when it landed looked something like a section of the Maine coastline. Then we waxed the thread, one of us holding the wax while the other drew the thread across it. Sure enough, it now landed in a smooth and gentle curve, like the shoreline of a smooth New Jersey beach.

The story of the Stoppages not only teaches us that the interaction between science and art in the early twentieth century is more extensive than we commonly believe, but also the promises and perils of investigating this connection.[9]

The Stoppages parodies—it playfully imitates characteristic features out of context—the global metrological network. Instead of a stiff alloy, a thread. Instead of an invariant, a fluttery filament. Instead of stamping out disturbances and contingencies, celebrating them. Instead of a straight line curving in non-Euclidean space, a thread twists in an artist's studio. Instead of an artifact morphing into a standard by a declaration by a network of international scientific institutions, a fiber becomes a standard by declaration of an artist and a network of art institutions. Instead of the metric system being liberating to humanity—as its institutors, the French revolutionaries, saw it—it is a parody of the metric system that is liberating to an artist.

Duchamp's *3 stoppages étalon* does all this sparely and in a pleasurable and whimsical way—the instant one turns literal-minded one ceases to "get" it—that generates a satisfyingly endless play of thoughts about conventions and standards. What more could one hope for from art?

MOCKTROLOGY

Making fun of units and measures is the subject of a science that might be called *mocktrology*. One of the most famous examples—rarely neglected in books about measurement—concerns the Smoot. In 1958, as part of a fraternity stunt, the Harvard Bridge was measured by a group of students who selected one of their members, Oliver R. Smoot (5-foot-7), as the ruler; the result was 364.4 Smoots "plus or minus one ear." This now-famous measure has been memorialized by a plaque on the bridge, and in Google Calculator.

Like Duchamp's Stoppages, this is not entirely a joke. It illustrates an instructive classroom exercise about measures: you can use anything to measure, as long as it is willing, able, and convenient. I often ask classes to measure the volume of the classroom in a unit taken from inside the classroom itself, giving them the freedom to choose whatever they want. After debating possibilities—tables are usually rejected as solid but too heavy to move; pieces of chalk are easy to move but require too many repetitions—the students usually decide to use either someone's arm span or height so that he or she could move him- or herself. The first time that I did this, the students chose a certain Tina, because her height seemed to be about half the ceiling size (she seemed the only measure in the class that would eliminate the need for fractions in that dimension) and she was willing to lie down on the floor and then move herself repeatedly. The room proved to be 104½ cubic Tinas.

An Austrian physicist told me that an important measure in the region where he grew up was the spaetzlehobel, or the size of the holes in the sieve through which dough is forced to make spaetzle.

Humorous fictional units have their own Wikipedia page and increase in number every year. The classic example of a unit to measure a subjective property is the helen, after the line in Christopher Marlowe's *Doctor Faustus* referring to Helen of Troy's face as having "launch'd a thousand ships," implying the millihelen as the amount needed to launch one. One physicist wrote to me that his favorite units were the "cows" and "dogs" that his physics teacher wrote as axis titles on graphs when he had omitted them as a student. The sheppey, invented by Douglas Adams and John Lloyd, is the closest unit of distance at which sheep remain picturesque, or about 7/8 mile; the warhol is a measure of fame equal to 15 minutes' worth; this makes a megawarhol equal to 15 million minutes, or about 28½ years. The beard-second, or the length that a "standard beard grows in a second," is a jocular body measure inspired by and created in form parallel to the light-year, or the distance that light travels in a year.[10]

One group of humorous units names the "least amount." A retired instrument engineer wrote to me that a midges was "the smallest amount of linear or rotational movement achievable at the output of a given mechanical or electrical device, within the constraints of static friction, the adjustment mechanisms provided by the designer, and the dexterity of the operator." Other engineers wrote about the gnat's whisker, a subdivision of that somewhat larger, well-known unit the "gnat's ass" or "cock hair" (in many British regional accents this is "cock 'air"), as in "Move that needle a cock 'air to the left!" A physicist and software engineer explained to me the technical terms used to capture signals: a "tweak" is a fine-tune; a "twiddle," a gross manipulation; a "frob," an aimless manipulation.

Even unit prefixes are joke topics. Official SI prefixes march upward by thousands, with the high end represented by peta, or 10^{15}; exa, or 10^{18}; zeta, or 10^{21}; and yotta, or 10^{24}. In 2010, a physics student from the University of California, Davis, created a Facebook petition to lobby for yet another prefix, 10^{27}, to be named "hella," California slang for "a lot of." A stampede of other suggestions followed, with letters to the international science magazine *Physics World* proposing to extend units downward to 10^{-27} with the "tini"–pronounced "tee-nee"–followed by the "insi" ("eensy") for 10^{-30}, and "winsi" ("weensey") for 10^{-33}. Another correspondent noted that the SI prefix "zepto" for 10^{-21} was clearly a mistake for "zeppo," and proposed that the prefixes for 10^{-27}, 10^{-30}, and 10^{-33} should be "groucho," "chico," and "harpo," respectively.[11]

A retired physics professor wrote to me that his son, a computer software engineer, is an SI fundamentalist who refuses to observe normal birthdays and measures his age in megaseconds—though the second is an astronomical accident, at least it's SI and now defined without reference to the Solar System. He celebrates the passing of time every 50 megaseconds. His father has taken on the spirit, if not the letter, of his son's position by informing students that the length of each lecture is a microcentury (52 minutes, 36 seconds).

Units serve human needs, and the needs of everyday life are diverse and continually changing. Humorous ones have a function of their own in that they satirize, spoof, or symbolize how arbitrary the process of measure taking can be. The way we measure is usually taken for granted and tends to remain in the background of ordinary life; our measurement system tends to become noticeable (except to those who are in charge of it) only in the event of a breakdown. The most notorious episode of this occurred in 1999, when a $125 million Mars spacecraft crashed after its two different teams of engineers used two different systems, imperial and metric, to program its rockets. Airplanes have occasionally been forced to land because of a miscommunication, traced to a similar cause, about the amount of fuel put into them. So it seems that satirical and silly units have an indisputable cultural value—an irony in and of itself—for they remind us of the conventional character of units in a humorous rather than a disastrous way.

CHAPTER NINE

DREAMS OF A FINAL STANDARD

The non-scientific mind has the most ridiculous ideas of the precision of laboratory work, and would be much surprised to learn that, excepting electrical measurements, the bulk of it does not exceed the precision of an upholsterer who comes to measure a window for a pair of curtains.

—CHARLES S. PEIRCE

Charles Sanders Peirce may have been exaggerating somewhat. But when it came to metrological matters, Peirce (pronounced *purse*) knew what he was talking about. Peirce (1839–1914), one of the most remarkable, eccentric, and unheralded American-born geniuses, was a logician, scientist, mathematician, and America's most original philosopher, a founder of the philosophical school known as pragmatism. He was also one of America's most important metrologists.[1] He made precision measurements, and improved techniques for making them. His work helped remove American metrology from under the British shadow and put American metrology on its feet.

Peirce was the first to experimentally tie a unit, the meter, to a natural standard, the wavelength of a spectral line. A natural standard had been the dream of many scientists in several countries for centuries. The French had tried to tether it to the earth's dimensions, and the British to a seconds pendulum. Their attempts failed; Peirce was the first finally to show how it could be achieved.

Amazingly, his contribution has not received much attention, for several reasons. For one thing, though many scientists realized the importance of his feat at the time, he never finished it to his satisfaction, leaving fragmentary reports about it in his 12,000 published pages and 80,000 pages of handwritten notes and letters, mostly on logic, mathematics, science, and philosophy. More significantly, Peirce's idea—counting wavelengths—was almost immediately taken up and improved by Albert A. Michelson, a far more well-known American scientist who used a superior technique. In 1907, Michelson became the first American to win the Nobel Prize in the sciences, for a famous experiment demonstrating the nonexistence of the ether; credit for the first experiment that connected the meter and wavelengths of light is usually attributed to him instead of Peirce. Finally, Peirce's chaotic professional and personal life hindered a comprehensive assessment of his contributions.

On the professional side, Peirce was a prolific and perpetually overextended polymath. He scarcely began one ambitious project before starting others, and rarely finished them. On the personal side, he suffered from several illnesses, including a severe inflammation of facial nerves (now called trigeminal neuralgia), and extreme mood swings that today would be diagnosed as bipolar disorder. The common treatments of the day were ether, opium, and cocaine, which compounded the social and physical challenges created by the diseases. His symptoms and difficult temper worsened as he aged. Peirce nearly landed professorships at Harvard and Johns Hopkins, but was rude and aggressive even with supporters and so wrecked his excellent chances. His fractious personality is so difficult to encompass in a short description that Joseph Brent, the author of the only full-length biography of Peirce (1993), threw up his hands and resorted to a laundry list of psychopathologies:

> On the manic side he exhibited driven paranoid, and impulsive actions; extreme insomnia; manic grandiosity and visionary expansiveness; hypersexuality; extraordinary energy; and irrational financial dealings, including compulsive extravagance and disastrous investments. On the depressive side, he exhibited severely melancholic or depressive states characterized by suicidal feelings or flatness of mood, which were accompanied by inertness of mind, inability to feel emotion, and an unbearable sense of futility.[2]

Some future biographer may be able to integrate such behaviors with Peirce's life of multitasking, extraordinary productivity, and original insights to produce a more comprehensive and positive portrait of this

CHARLES SANDERS PEIRCE.

remarkable American genius. Peirce indeed had an original personality, but he had an even more original mind.

PEIRCE'S BACKGROUND

Peirce had an enviable start. He was born into an elite family in Cambridge, Massachusetts. He was handsome and articulate and collected friends in high places, but then spent his career squandering those advantages. Peirce got fired from every serious job he had, quarreled with admirers and benefactors, engaged in behavior that peers found scandalous, and wound up in abject poverty.

Charles's father Benjamin (1809–1880) was professor of mathematics and astronomy at Harvard and an administrator in the U.S. Coast Survey. Grooming his son for science, Benjamin sent Charles to private school and Harvard, from which the boy graduated in 1859 at the age of nineteen. Charles attended the Lawrence Scientific School, Harvard's early graduate program in engineering and the sciences, which Benjamin had helped to found, and received a master of science degree in chemistry, summa cum laude, in 1863. Shortly after, in Cambridge, he met and married Melusina Fay Peirce, who would become a prominent early American feminist activist and author. "Zina," as she was known, was also deeply religious, and thought that adultery was a crime whose penalty ought to be life imprisonment or death. Had this indeed been the law of the land, she would have been an early widow, because to Zina's fury Charles, despite being a workaholic and despite his diseases and addictions, carried on numerous affairs.

As a youth, Charles was fascinated by logic, viewed himself as a logician until the end of his life, and yearned to work full time on the subject. But Benjamin, wishing to promote his son's continued education, arranged a series of scientific apprenticeships. One was with the

Coast Survey. In the 1860s it was still the preeminent scientific agency of the U.S. government; its recent work on the purchase of Alaska from Russia and the proposed purchases of Greenland and Iceland from Denmark gave it a high political profile. Charles's first official position at the Survey came in 1861, at the beginning of the Civil War, when his father needed a new computational helper after he lost one to the war; he got his son appointed as the replacement. In 1867, Benjamin became superintendent of the Survey after Alexander Bache, its previous superintendent, died. Benjamin ran the office from Cambridge and appointed his son as an aide and then, in 1872, as an assistant directly beneath him in rank. Zina, meanwhile, attempted to organize Massachusetts housewives for cooperative housekeeping and retail merchandise projects, promoting the idea in articles for the *Atlantic Monthly* and other magazines, and founded the Cooperative Housekeeping Association in Cambridge (1870). The time was not ripe; her efforts collapsed after adverse pressure from outraged husbands.

During the same period, Benjamin arranged a second apprenticeship for his son at the Harvard Observatory, whose director, Joseph Winlock, was a close friend. Spectroscopy, the study of light when separated into its constituent wavelengths or spectral lines, was in its infancy, and the observatory acquired its first spectroscope in 1867. Charles helped Winlock with observations of stellar wavelengths. As Winlock's assistant Peirce was one of the first to observe the spectrum of argon and was assigned to expeditions to carry out measurements of two solar eclipses. The first was in Kentucky in 1869; the second in Sicily in 1870, in an expedition whose measurements played a role in the theory of the corona. Peirce also embarked on an ambitious attempt to use the relative brightness of stars to determine the shape of the galaxy, and of its stellar distribution. In 1872 he wrote to his mother about his overstretched work habits and lack of sleep: "On clear nights I observe with the photometer; on cloudy nights I write my book on logic which the world has been so long & so anxiously expecting."[3]

The logic book, like many of Peirce's projects, was never completed. But his spectroscopic observations resulted in *Photometric Researches* (1878), the only book of his to be published in his lifetime. He wrote to his father that it would be *the* book on the subject. It might have been, had it been promptly published when his research ended in 1875. But quarrels with his Harvard bosses and his other commitments stalled publication until 1878, blunting the book's impact.

Also in 1872, Peirce became a founding member of the Metaphysical Club, a discussion group of prominent intellects of the day, whose members included the philosopher William James and the jurist Oliver Wendell Holmes, Jr.[4] In the following years, this group originated the basic concepts of pragmatism as a philosophical movement, according to which meaning and truth are sought in the practical consequences of conceptions and beliefs. As James once put it: "Truth is what works."

The year 1872 was a banner year for Peirce. He had two high-profile positions, excellent prospects for further advancement, and influential friends. He still considered himself a logician. But career prospects in logic were few, and he saw his scientific work as enriching his ideas about logic. Among other things, it left him with a better understanding of the importance of precision, and the difficulties of achieving it.

Two kinds of precision research in particular prepared Peirce for the measurement that tied the meter to optical wavelengths. One, from his observatory work, was photometrics, which involved the theory and practice of spectroscopes, diffraction gratings, and wavelength measurements. The other was experience in gravimetrics, which he was about to get at the Coast Survey: it involved the theory and practice of pendulums and their calibration.

A NATURAL LENGTH STANDARD

Metrology, as we saw in Chapter 6, advanced swiftly in the second phase of the Industrial Revolution, and with the events inaugurated by the 1875 Treaty of the Meter. Peirce's remark, cited at the beginning of this chapter, was accurate; electrical measurements were the vanguard. The expanding telegraph industry, particularly in Great Britain, drove the need for electrical standards and instruments to set and supervise them.[5] But that was just the beginning. Increasing practical application of electricity, and the process of electrification of entire regions, attracted both theoretical and practical attention, leading to numerous international gatherings at which scientists and engineers discussed electrical units and standards. In 1832 in Germany the mathematician and scientist Carl Friedrich Gauss (1777–1855) proposed an ingenious scheme to consolidate all units—even electrical ones—to three. A unit of speed, for instance, can be forged from units of distance and time, 60 miles per hour, say. A unit of force takes three units: what gets a certain mass moving a certain speed in a certain time. Gauss showed that magnetism and electricity do not require their own units, but can be measured using these three mechanical units: the amount that can exert a certain force on a certain mass at a certain distance. Gauss called this an "absolute system," meaning something final that did not require a "relative" measure in terms of another magnet or electrical source. Today, we would refer instead to "base" units. A few years later, working with physicist Wilhelm Weber (1804–1891), Gauss demonstrated that units of length, time, and mass, rather than units related to traditional magnetic instruments, could be simplified, consolidated, and extended to electrical and nonmechanical phenomena.

Following up this idea in the 1860s, James Clerk Maxwell and William Thomson (Lord Kelvin) developed the concept that a set of fundamental or "base" units could be coupled with affiliated or "derived" units in an efficient and "coherent" system—coherent in the sense used by metrologists, that conversion factors between units are not required. In 1874, taking this idea still further, the British Association for the Advancement of Science proposed a coherent unit scheme called the CGS (centimeter, gram, second) system. A few years later, the organization introduced what proved to be a superior coherent system for measuring electromagnetic phenomena, incorporating the ohm (resistance), volt (force), and ampere (current).

But the quest for precision in mechanical standards was not far behind, with international industrial expositions providing a strong stimulus for greater international cooperation in setting measurement units and standards.[6] The quest for a universal standard of length, too, was renewed.

The French scientist François Arago expressed the dream as follows: "A measure that can be reproduced even after earthquakes and terrifying cataclysms shatter our planet and destroy the standard prototypes preserved at the archives."[7] The success of Maxwell's comprehensive theory of electromagnetism in the 1860s convinced many scientists that the meter eventually could be tied to the wavelength of a spectral line. At the beginning of Maxwell's magnum opus, *Treatise on Electricity and Magnetism* (1873), he remarked on the failure of both the imperial and metric systems to connect the dimensions of units of length with natural units. New measurements, he pointed out, showed that the meter was not in fact the ten-millionth part of the arc of a meridian, and for all practical purposes is the length of a standard preserved in Paris. "[T]he mètre has not been altered to correspond with new and more accurate measurements of the earth, but the arc of the meridian is estimated in terms of the original mètre." Maxwell continued, "in the present state of science the most universal standard of length which we could assume would be the wave length in vacuum of a particular kind of light, emitted by some widely diffused substance such as sodium, which has well-defined lines in its spectrum."

In his characteristically wry manner, Maxwell then expressed the desire for a universal standard of the time: It "would be independent of any changes in the dimensions of the earth, and should be adopted by those who expect their writings to be more permanent than that body."[8]

Peirce's route to seeking a natural standard began in 1872, when Hilgard, the head of the Coast Survey's Washington office, left for Paris to represent the United States at the International Metric Commission. In Hilgard's absence, Superintendent Peirce appointed his son Charles to be acting head. One of Peirce's duties was to direct the Office of Weights and Measures, which then belonged to the Survey's Washington bureau. This brought Peirce into contact with the growing international metrological community.

Benjamin, meanwhile, continued to elevate the Coast Survey's profile. In 1871, he persuaded Congress to mandate a transcontinental geodetic survey along the thirty-ninth parallel, connecting surveys already taken on both coasts. Geodesy was replacing surveying as the Survey's chief work, and in 1878 the agency would be renamed the Coast and Geodetic Survey. Pendulums were the principal gravimetric instruments of geodesy, and after Hilgard's return Benjamin put Charles in charge of the Survey's pendulum research. This work involved the use of precision length standards to calibrate the pendulums.

The International Geodetic Association (IGA) had organized a network of gravimetric surveys, and by 1872 had chosen as its instrument a reversible pendulum, one formed by a solid rod that was swung first with one end up and then the other end up. If a reversible pendulum swings with equal periods, the distance between the two knife edges is equal to the length of an ideal or simple pendulum of the same period. This fact allowed pendulum users to ignore most of the disturbing factors, and thereby transformed the pendulum into a valuable, highly sensitive scientific instrument, able to measure any factor that disturbed the pendulum's simple motion. The IGA reversible pendulum had been designed by the German astronomer Friedrich Bessel and manufactured by A. & G. Repsold and Sons, a Hamburg instrument maker. Peirce ordered a Repsold pendulum for his own use, but completion was delayed while the company—which though highly specialized was also sought-after and extremely busy—filled orders for instruments sought by astronomers to measure the 1874 transit of Venus across the sun's disk as seen from earth. Such transits only occur once or twice a century, and astronomers and instrument makers put aside other business to prepare for this phenomenon. When Peirce's order was finally completed in 1875, he traveled to Europe to pick it up. During this trip, Peirce met several prominent scientists and scholars: he impressed some of them with his mathematical and logical proficiency, and he put off others with his increasingly unsocial and odd behavior. He met Maxwell, with whom he discussed pendulum theory at the then-new Cavendish Laboratory. He also met the novelist Henry James, who wrote to his brother William that Peirce had "too little art of making himself agreeable."[9]

Pendulums appeared to involve little new theory. Still, Peirce found a way to be original, applying his logical and mathematical knowledge to analyze systematic errors due to the pendulum's mount. He developed a theory of such effects, showed that they explained various measurement discrepancies, and designed an improved instrument.[10] Four identical instruments of Peirce's design were later constructed by the Survey, one of which is now housed at the Smithsonian Institution.

Peirce was asked to report his findings to the Special Committee on the Pendulum of the IGA, which met in Paris while he was in Europe. That made him the first American scientist invited to participate in the

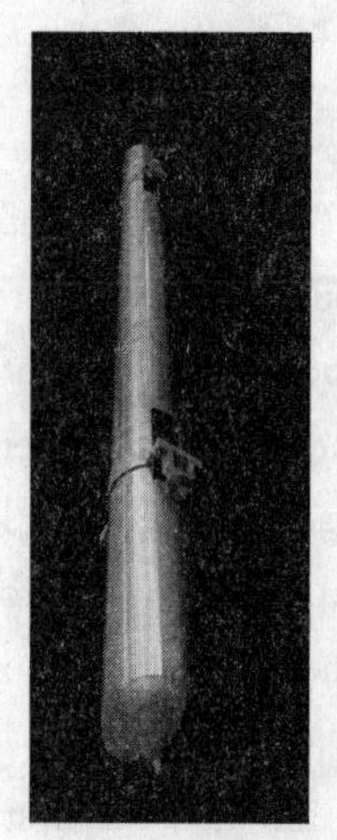

A GRAVIMETRIC PENDULUM, ONE OF SEVERAL DESIGNED AND USED BY CHARLES S. PEIRCE IN THE 1870S AND 1880S.

policy meeting of an international scientific association.[11] He returned to the United States in August 1876, and brought back a brass meter standard for calibrating American standards. The standards were numbered, and the one he brought was "no. 49."

After his return, Peirce suffered a major nervous breakdown—the first of more than half a dozen. The pain from his facial neuralgia was intense, his bipolar disorder worsened, and his behavior and moods were increasingly erratic. Still, he moved to New York City to work simultaneously on gravimetrics and spectroscopy at the Stevens Institute of Technology in nearby Hoboken, New Jersey. Zina remained in Cambridge, effectively ending their relationship, if not yet the marriage; she had not tolerated well his affairs and obsessive behaviors. The two had no children. Peirce took up with Juliette Pourtalai, an exotic and mysterious woman about whom little is known (she occasionally but implausibly claimed to be a princess). Peirce conducted the affair "far less discretely than the times demanded," appearing with her at important public events.[12]

Nonetheless, Peirce continued to be prolific. His gravimetric work resulted in a lengthy paper entitled "Measurements of Gravity at Initial Stations in America and Europe," which is "one of the classics of geodesy and the first notable American contribution to gravity research."[13] It left Peirce with a strong sense of the importance of the international nature of science. As he wrote, "The value of gravity-determinations depends upon their being bound together, each with all the others which have been made anywhere upon the earth . . . Geodesy is the one science the successful prosecution of which absolutely depends upon international solidarity."[14]

Peirce's spectroscopic work at Stevens initiated the first measurement to tie the meter—or any unit—to a wavelength. The idea had been mentioned before—by Babinet in 1827, and by Maxwell and other British scientists in the early 1870s. But realizing the idea in practice was another matter. Peirce was the first to do it.

The principle was simple, involving two measurements. One was to determine the angle of deviation of a ray of light passing through a diffraction-grating, the other to establish the spacing of the grating lines. The relation, well known to physicists, among the spacing of the lines, the wavelength of the light, and its angular deviation then would connect the wavelength with the meter.

Peirce was motivated by the well-known vulnerability and uncertainty of artifact standards, whose dimensions had been known to change over the years. He proposed that "the standard length may be compared with that of a wave of light identified by a line in the solar spectrum."[15] The proposal was not without problems. It involved "the assumption that the wave-lengths of light are of a constant value," Peirce wrote in 1879. Peirce and all other scientists at the time assumed that, just as sound waves travel in the medium of air and water waves in water, so light waves must have a medium called the ether. In the same way that the velocity and wavelength of sound and water waves are affected by the motion of the air and water they pass through, so the velocity and wavelength of light must be affected by the motion of the earth in the ether. In 1881, and again in 1887, Albert Michelson and Edward Morley would attempt to detect evidence of this motion using a device called an interferometer, which splits a beam of light, bounces the two beams off mirrors in two different directions and recombines them, and the instrument can then detect minute differences in velocity or wavelength. When they failed to detect any such difference, the unexpected result would shock the scientific world. But Peirce's remark predated that result, and he was rightly worried about possible ether effects: "[T]here may be a variation in wave-lengths if the ether of space, through which the solar system is travelling, has different degrees of density. But as yet we are not informed of such variation."[16] Nevertheless, Peirce's idea was not hopeless; if there was such a variation in wavelengths due to the motion of light in the ether, connecting a length standard and wavelengths of light would have to include a correction factor involving the direction and velocity of the ether "wind."

Peirce worked on this project for several years after his arrival at Stevens, though his labors were interspersed, as always, with setbacks due to illness and interruptions due to overcommitment. The idea was based on the relation

$$n\lambda = d \sin \theta$$

between the wavelength λ of a line, the line spacing d of the diffraction grating, the bending angle θ, and the order n of the diffraction pattern.

Diffraction gratings are optical tools that consist of a set of parallel lines finely cut into glass or metal. Light falling on such gratings splits and diffracts, with different beams traveling in different directions. An early version had been produced by Thomas Jefferson's friend David Rittenhouse, who in 1785 manufactured an instrument of fine, parallel hairs, 106 to an inch, but who did not recognize the far-reaching potential of the device. Four decades later Frauenhofer began to explore this potential with gratings made out of wire, and of fine lines scratched on glass. Following the development of spectroscopy, which made it possible to examine the characteristics of starlight, diffraction gratings had become indispensable instruments, replacing prisms as precision instruments in spectroscopy and optics. "No single tool," wrote MIT Science Dean and diffraction grating pioneer George R. Harrison, "has contributed more to the progress of modern physics than the diffraction grating."[17]

Peirce, like others, realized that if the lines could be drawn finely enough, a connection could be made between the wavelength of a spectral line being diffracted and the spacing of the grating that was diffracting it, allowing that wavelength to be turned into a length standard. British astronomers talked about this prospect in the early 1870s.[18] Success would depend on the quality of the grating.

During that decade, Lewis M. Rutherfurd (1816–1892), another scientist working partly at Stevens, made the best gratings. An independently wealthy amateur astronomer and instrument maker, Rutherfurd had constructed an observatory in the garden of his home at the corner of Eleventh Street and Second Avenue in New York City.[19] He built a micrometer to measure his solar photographs; Peirce used this micrometer to calibrate centimeter scales for his pendulums. Rutherfurd became interested in spectroscopy when Robert Bunsen and Gustav Kirchoff made their startling assertion in 1859 that spectra were fingerprints of chemical elements. Before turning to diffraction gratings, Rutherfurd used prisms. In 1867, faced with the task of ruling gratings before electric motors were available, he built an ingenious machine to

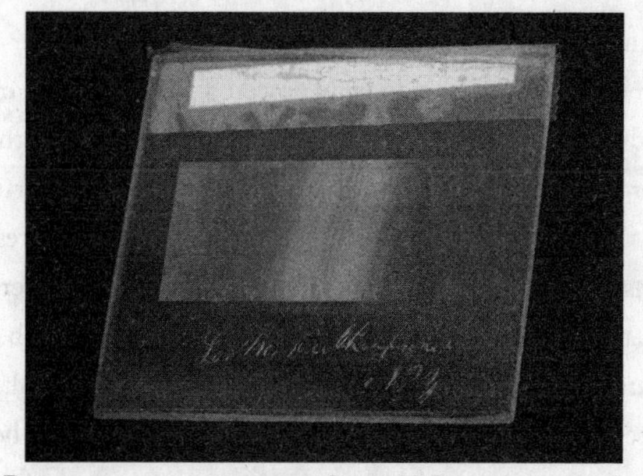

Diffraction grating made by Lewis M. Rutherfurd in 1872 with the ruling engine shown below. Rutherfurd's signature appears on the frame, and the line spacing, "12960 to the inch," is written on the attached tape.

Engine built by Lewis M. Rutherfurd in the 1860s to rule lines for diffraction gratings. Powered by a water turbine that ran off the New York City water system, the engine's wheels and gears continually moved the grating's platform (*e*) under the diamond-tipped stylus (*m*) that scribed the lines. After a line was completed, the micrometer screw (*d*) advanced the substrate laterally by precisely the desired line spacing.

rule gratings on blanks made of glass or speculum, a copper-tin alloy. The engine was driven by a turbine powered by water from city pipes. It used a diamond stylus and a micrometer screw to advance the blanks.

Rutherfurd treated his gratings with great care, as an artist would treat the composition of a painting. Several have been preserved at the Smithsonian. About 4 centimeters wide, the gratings were usually signed, dated, and inscribed with information about the number of lines per inch and kept in carved laquer boxes of the sort made for daguerreotypes. Because Rutherfurd gave away the gratings at cost, he became quite popular among spectroscopists.

Peirce had used the micrometer to calibrate centimeter scales for his pendulums and greatly appreciated the quality of Rutherfurd's work. When Peirce came calling for a grating, Rutherfurd's engine had a wheel with 360 teeth on its circumference able to rule 6808 lines per centimeter, and Rutherfurd gave him one. Applying his usual care, Peirce noted imperfections in Rutherfurd's grating: the stylus left a burr on one side of each line, which he found a way to remove, thereby improving the precision he could obtain with it. Rutherfurd's work, Peirce wrote, at last made it practical to think about measuring "a wave length to one-millionth part of its own length." Peirce considered Rutherfurd's contribution so essential that one unpublished manuscript lists him as a coauthor.

By 1877, however, Rutherfurd was ill and had to curtail work. Peirce, meanwhile, quarreled with Stevens colleagues and with his Survey bosses, threatening to quit, but was wooed back by Carlile Patterson, who had replaced Charles's father as superintendent three years previously. Then, in September 1877, Peirce left for Europe for two months to attend sessions of the fifth general conference of the IGA. His talk there was the first formal presentation by a U.S. scientific representative at a formal meeting of an international scientific association.

En route, on shipboard, isolated from troubles, Peirce wrote an essay on scientific method, "How to Make our Ideas Clear," and he translated "The Fixation of Belief," which he had originally written in French. Those essays, plus another four written later and published as "Illustrations of the Logic of Science," are seminal essays on Peirce's thought. He articulates the basic principles of the early phase of his pragmatic philosophy, in which it was a logic of science. These essays reveal the influence of his scientific work and his metrological experiences in particular, which allowed him to appreciate features of science that eluded—and still elude—those with a more formalistic outlook. In fact, a reader familiar with Peirce's scientific work might be tempted to call these essays the meditations of a metrologist.

Just as a measurement system, which creates a platform for carrying out both ordinary and scientific activities, is something we take for granted and only draws our attention when it breaks down or fails to provide what we need, so Peirce says that in ordinary life we have habits that give us comfort and security, and which provide a platform for action, as long as we do not have to think about them. But just as a measurement system is never perfect and does not anticipate everything human beings will ever demand of it, so our beliefs never fully mesh, seamlessly, with the world. This gives rise to anxiety and dissatisfaction, or what Peirce calls the "irritation of doubt." Peirce then describes four ways of overcoming this irritation: tenacity (stubbornly rejecting the reality of what is causing the irritation), authority (using an institution such as the state to impose a solution), the a priori method (searching for some purely rational beginning point (which turns out to be a form of tenacity), and the scientific method (turning away from oneself to collaborate with the world, inquiring into nature to arrive at a solution).

In conducting inquiry, Peirce continues, scientists inherit the often-defective tools, hypotheses, and experiences of predecessors. Yet it does not matter if these are imperfect, because science is a fallible process in which a community of inquirers corrects errors in ongoing revision. Knowledge grows, not in a staccato-like way in which one representation replaces another, but in a continuously expanding process in which a concept's meaning is not an abstraction or picture, but the totality of its effects on the world.

In those essays, Peirce's pragmatism is different from that of his friend William James. Peirce approached science as a matter not of solitary scholars confronting puzzles in private, as James had, but of networks of competent people working in networks of labs in an inherently public enterprise. Peirce also valued what he called the "economy of research"; that an important part of science is maximizing resources—"money, time, thought, and energy"—when deciding what to work on. He realized there could never be absolute precision. "Dealing as they do with matters of measurement, [physicists] hardly conceive it possible that the absolute truth should ever be reached, and therefore instead of asking whether a proposition is true or false, they ask how great its error is."[20]

On his return to Stevens, in late 1877, Peirce continued his work tying the meter to light. He measured the angular displacement of the image of a slit by a grating. Then, using a device called a "comparator" that he had built himself, he compared the grating's line spacing with the units of a glass decimeter that he had calibrated using his no. 49 standard meter. He was, in effect, calibrating the grating spacing in units of wavelength.

Peirce now found that the appearance of "ghosts," faint lines that appear on either side of principal spectral lines, retarded greater resolution. These lines were clearly unreal, an artifact created by the instrument rather than a phenomenon of nature, because they only appeared in spectra created by gratings and never by prisms. Ghosts were created by tiny imperfections in the micrometer screws that had ruled the grating. But as he had done with pendulums, Peirce treated the imperfections as opportunities: he measured them, developed a theory, and applied it to correct the measurements.

With these adjustments, Peirce tried measuring a spectral line produced by sodium. He chose the line because it was easy to produce and relatively sharp. His idea would involve bootstrapping a standard: if the wavelength of a spectral line could be measured in meters precisely enough, it opened the door to redefining the meter in terms of that wavelength.

But he again ran into several sources of error, among them the thermal expansion coefficient of the glass grating and the quality of the thermometer used to measure its temperature. Peirce published a brief progress report, "Note on the Progress of Experiments for Comparing a Wave-length with a Metre," in the July 1879 issue of *American Journal of Science*. "As soon as that"—reduction of the errors, and various calibrations—"is done *a meter* will have been compared with a wave length." This brief and unpretentious "note" is the key published source for Peirce's revolutionary work.

Meanwhile, Peirce was proposed for a chair in physics at Johns Hopkins University. He had made enough personal enemies that he did not get the position, but was invited to lecture in logic. Even his best students—among them John Dewey, soon to be a fellow pragmatist philosopher—found him difficult to understand. But they discovered he was dazzlingly facile. For instance, he could simultaneously write out a mathematical problem with his right hand on the blackboard, and its solution with the left. Yet Peirce continued to infuriate others, initiating a particularly ugly quarrel with a visiting math professor over priority for a mathematics discovery.

Peirce kept improving his wavelength measurement, wrote a brief report in *Nature* (1881) entitled "Width of Mr. Rutherfurd's Rulings," made a report to the U.S. superintendent of Weights and Measures on the work, and began a summary, "Comparison of the Metre with a Wave-Length of Light." But this summary, like so much of Peirce's other work, remained unpublished.

His personal life, always tumultuous, was beginning to unravel. Until now, his father Benjamin, or some other supporter, had usually rescued him when he wound up in personal or financial difficulty. But his two foremost protectors died: his father Benjamin in 1880, and Superintendent Patterson the following year. For years Benjamin had tried to groom Charles to become superintendent of the Survey, but his son's erratic behavior and cantankerous personality shipwrecked that plan. Thus Hilgard, who was incompetent, uninterested in research, impatient with overcommitted people like Peirce, and himself in declining health succeeded Patterson.

Peirce finally divorced Zina officially in 1883 and married Juliette a few days later. Though he and Zina had been separated for 7 years, colleagues found their haste scandalous. Peirce fought with his cook, who sued him for assault with a brick. "I have lately been offending people everywhere," he wrote to Hopkins President Gilman in 1883.[21] Unfortunately, the offended parties included the Hopkins trustees, who dismissed Peirce in 1884.

For a few years, Peirce worked with Hopkins students outside the classroom and kept his assistantship at the Survey. From October 1884 to February 1885 he headed the Survey's Office of Weights and Measures. But Hilgard's own health and behavior grew worse, and charges were lodged against him of drunkenness and other forms of misconduct. Peirce became ensnared in the ensuing scandal—for once not of his own making—and subjected to congressional investigation with the rest of the office. He was forced to resign along with Hilgard. Suddenly, he had no regular position to count on.

MICHELSON AND MORLEY

Peirce's attempt to use a wavelength of light as natural length standard inspired others. One was at Hopkins, where Henry Rowland (1848–1901)—Peirce's former rival as candidate for chairman of the physics department—began to manufacture gratings superior to Rutherfurd's. Rowland's work helped make the university a center of optical research from the 1880s to the Second World War.[22] His student Louis Bell achieved an accuracy of 1 part in 200,000 with Rowland's gratings.[23]

Another attempt was made at the Case School of Applied Science in Cleveland, Ohio, where Albert Michelson had read Peirce's publications. Michelson realized that the interferometer which he and Edward Morley had developed and were just then using in their attempt to detect the drift of the ether could also be used to make precise wavelength measurements of the sort Peirce was attempting.

In June 1887, after getting initial results in their experiment on the speed of light, Michelson and Morley conducted preliminary measurements. Their paper "On a Method of Making the Wave-length of Sodium Light the Actual and Practical Standard of Length" begins, "The first actual attempt to make the wave-length of sodium light a standard of length was made by Peirce."[24] But, they pointed out, Peirce's measurements, "which have not as yet been published" (and never would be) had many systematic errors.

The Michelson-Morley interferometer, which split a beam, sent its two parts along different paths at the end of which they bounced off a mirror, and recombined them to create an interference pattern, overcame these errors. When the beams recombine, they form an interference pattern, a series of fringes, dark and light patterns, each one a wavelength. By varying the mirrors slightly—the length of the bounce—you can get the fringes—the wavelengths—to shift. It thus

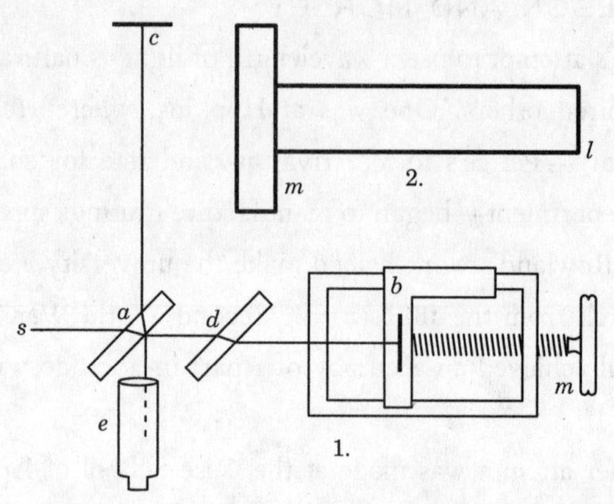

INTERFEROMETER USED BY ALBERT MICHELSON AND EDWARD MORLEY TO MEASURE THE WAVELENGTH OF LIGHT. A BEAM OF LIGHT FROM A SODIUM LAMP (*s*) FALLS ON A PLANE OF GLASS (*a*) THAT SPLITS THE BEAM, SENDING PART TO A MIRROR AT *c* AND PART TO A MIRROR AT *b*. BOTH BEAMS BOUNCE BACK TO *a* WHERE THEY RECOMBINE AND ARE REFLECTED INTO A TELESCOPE AT *e*. IF DISTANCES *ac* AND *ab* ARE EXACTLY EQUAL, AN OBSERVER AT *e* SEES BLACK BECAUSE OF THE INTERFERENCE BETWEEN THE TWO BEAMS. MOVING THE MIRROR (*b*) WITH A MICROMETER SCREW (*m*) PRODUCES A SERIES OF LIGHT AND DARK ALTERNATIONS, WHOSE NUMBER IS EXACTLY TWICE THE NUMBER OF WAVELENGTHS IN THE DISTANCE THE MIRROR HAS MOVED. "[T]HUS," MICHELSON AND MORLEY CONCLUDE, "THE DETERMINATION CONSISTS ABSOLUTELY OF A MEASUREMENT OF A LENGTH AND THE COUNTING OF A NUMBER" (*American Journal of Science* 34, 1887, P. 428).

provided a direct connection between wavelengths of light, the fringes, and distance, the slight movements of the mirror. At one mirror Michelson and Morley installed a micrometer to move the mirror a precise distance while counting interference fringes—light-and-dark alterations, each set a wavelength. By using a micrometer and fringes where Peirce had simply used a ruler, they greatly reduced measurement errors. In effect, they used the wavelengths, the light-and-dark alterations, as their ruler. Their work dramatically illustrated the fundamental limitations of Peirce's approach but also its revolutionary potential.

In 1887, Peirce—now almost 50—and Juliette moved to Milford, Pennsylvania, where they purchased an estate he named Arisbe after an ancient Greek town. This was not a retirement; it was another of Peirce's harebrained schemes. By then he was poor and in debt. He moved to the area because it was affluent and home to many prominent people. Peirce imagined that he could become a kind of guru for the neighbors and envisioned Arisbe becoming "a sort of Casino for fashionable people of 'cultural' tendencies to spend the summer, have a good time, and take a mild dose of philosophy."[25] This project, typically for him, did not work out as planned.

By this time, Peirce left most of his scientific work behind, though he occasionally dabbled in metrology and briefly considered returning to measuring wavelengths. He did not attend any of the metrological conferences becoming regular at the BIPM, outside Paris.[26] He had developed from a logician and laboratory-minded philosopher of science into an original thinker who incorporated insights about science into a comprehensive theory of inquiry; that theory included an appreciation for the roles of chaos and chance, with which he was familiar through metrology and thermodynamics. Nature is stochastic all the way down. Even in the most advanced inquiry, at some level scientists are still ultimately in the position of upholsterers cutting curtains for windows.

Though Peirce matured as an original philosopher, he never mastered the art of making himself agreeable. Always prolific, increasingly eccentric, he grew further estranged from friends and opportunities for livelihood. In 1897, he wrote to James that "a new world of which I knew nothing, and of which I cannot find that anybody who has ever written has really known much, has been disclosed to me, the world of misery."[27] In 1891, he was forced to resign from the Coast and Geodetic Survey. He was so destitute that William James appealed to friends on his behalf. In 1897 James wrote to James Cattell, the publisher of *Science*, "Glad to receive $10 for Peirce, who has few friends."[28]

In 1899 Peirce sought unsuccessfully to become inspector of standards for the Survey's Office of Weights and Measures. He had to watch from a distance when the National Bureau of Standards, for which he had agitated, was created in 1901. The fund established by James supported Peirce for the rest of his life. He died at Arisbe in 1914.

Between the Great Exposition of 1851 and the Treaty of the Meter in 1875, geodesy, engineering, and industrial concerns had been the principal drivers for consolidation and improvements in systems of weights and measures. Now there was emerging a new, scientific pressure. "Nearly all the grandest discoveries of science," wrote the electrical engineer and industrialist Werner von Siemens in 1876, "have been but the rewards of accurate measurements and patient long-continued labour in the minute sifting of numerical results."[29] More and more, the future of science and industry would depend on high-precision measurements.

Nations that had formerly rejected the metric system began to reconsider. Great Britain, which signed the Treaty in 1884, considered a bill to require adopting the metric system in 1896. The time seemed propitious: Britain was now trading with many countries that used the metric system—all European countries were now metric except for Britain and Russia—and meeting with stiff competition from Germany. Once again the attempt was unsuccessful, though the metric system was made legal but not mandatory in 1897.

Having a robust system of weights and measures had grown so essential to national and international commerce that several countries established laboratories for that sole purpose. The first was the Physikalisch-Technische Reichsanstalt in Berlin, Germany, whose driving force was Siemens and whose first head was Hermann Helmholtz. This was followed by the National Physical Laboratory in Great Britain, outside London, in 1900, and the National Bureau of Standards (NBS) in Washington, DC, in 1901.

In 1883, in a speech to the Institution of Civil Engineers, Lord Kelvin observed, "In physical science the first essential step in the direction of learning any subject is to find principles of numerical reckoning and practicable methods for measuring some quality connected with it. I often say that when you can measure what you are speaking about, and express it in numbers, you know something about it; but when you cannot measure it, when you cannot express it in numbers, your knowledge is of a meager and unsatisfactory kind; it may be the beginning of knowledge, but you have scarcely in your thoughts advanced to the state of Science, whatever the matter may be."[30]

Albert Michelson's later comment that "our future discoveries must be looked for in the sixth place of decimals," is only the most quoted of similar remarks made by scientists of the late nineteenth century.[31] These remarks are often interpreted as naïve lament: the era of great scientific breakthroughs is over, our picture of nature is nearing completion, and remaining scientific work is to be dry number-chasing. But the situation was far more complex. Not only were practical demands for precision mounting, but many researchers found the pursuit of precision exciting—even, in Victorian culture, an ethical good—and thought that the pursuit of precision might lead to previously inaccessible truths.[32]

This sentiment merely expresses the culmination of the implications of the world that Newton had discovered over 300 years previously. The world is open, infinite in many directions. Space, for instance, is both infinitely large and infinitely small. Measurements are never once and for all, but always an infinite task for scientists; any measurement can always be improved, and discoveries might just as readily occur on smaller scales as on bigger ones. Likewise, standards are not static, never definitions, but representations of something infinite, merely provisional drafts certain to be corrected, stand-ins for better ones to come. Before, the motivation for improved precision in weights and measures was driven by commerce and industry; now, it was self-perpetuating. In the second half of the nineteenth century, the quest for precision held practical urgency, national interest, military application, theoretical significance, and moral value.

Peirce's scientific work, and Michelson's improvement, reignited dreams of a natural standard. In 1887, after Michelson and Morley published their result, William Harkness, president of the Philosophical Society of Washington, expressed the dream as follows: Imagine in the distant future an interstellar traveler arriving at a distant planet, far beyond the realm of telescopes, who is asked to reproduce earth's standards of length, mass, and time after our sun has flared up and burnt the earth to a crisp. In the science of the seventeenth and eighteenth centuries, recovering earth's standards would be impossible, for everything on which we had based them would have vanished. "The spin of the earth which measures our days and nights," Harkness said, "would be irretrievably gone; our yards, our metres, our pounds, our kilogrammes would have tumbled with the earth into the ruins of the sun, and become part of the *débris* of the solar system. Could they be recovered from the dead past and live again?" For all science up until the past few years, the answer would be no. Recent developments have revived the dream. Because atoms are everywhere identical so is the light they emit, Harkness said, which our interstellar traveler can analyze with a spectroscope:

> By means of a diffraction grating and an accurate goniometer [an instrument to measure angles] he could recover the yard from the wave-length of sodium light with an error not exceeding one or two thousandths of an inch. Water is everywhere, and with his newly recovered yard he could measure a cubit foot of it, and thus recover the stan-

> dard of mass which we call a pound. The recovery of our standard of time would be more difficult; but even that could be accomplished with an error not exceeding half a minute in a day. . . . Thus all the units now used in transacting the world's business could be made to reappear . . . on the other side of an abyss of time and space before which the human mind shrinks back in dismay. The science of the eighteenth century sought to render itself immortal by basing its standard units upon the solid earth; but the science of the nineteenth century soars far beyond the solar system, and connects its units with the ultimate atoms which constitute the universe itself.[33]

Peirce had pioneered a path to a natural length standard; what of a natural weight standard? Here, too, Maxwell had had an idea. Molecules, he said in an essay, were all the same. He had theological grounds for doing so: God had manufactured them all alike, meaning they could be used as a standard if one could measure them precisely enough.[34] Indeed, as he explained in a later letter, "*if* we ever can measure the weight of a molecule of hydrogen, we shall *then* have a standard more permanent than any planet or heavenly body, but not till then, that is, not till what is now a merely conjectural estimate is converted into a well established physical constant. This molecular standard may possibly then be employed by those who expect that the authority of their numerical statements will outlast the planet on which they live."[35] If you could gather enough atoms, you would have a standard there as well. Atoms could supply two natural standards—the mass of one particular atom could potentially be used as a mass standard, and the wavelengths of the light emitted by electrons jumping between two different energy levels in a particular atom could provide the length standard.

In 1900 the latter natural standard, thanks to Peirce and Michelson, was on its way. The former was still a long way off.

CHAPTER TEN

UNIVERSAL SYSTEM: THE SI

Light, then—not the earth's meridian, nor the seconds pendulum—would provide the first natural standard. Plato compared light and its rays to the Good, for it nurtured and illuminated all that existed. Medieval scholars viewed it as a divine emanation or epiphany, the self-illumination of Being. It was the first thing that existed in the universe, according to the Bible ("Let there be light!"), and will be the last, according to those scientists who say that it will be what remains after all the matter and antimatter have annihilated themselves.

During the scientific revolution, light came to be viewed as a phenomenon just like any other, ubiquitous and unchanging, obeying mechanical principles. This inspired Babinet and other scientists, at the beginning of the nineteenth century, to propose that light might provide a natural standard. In the mid-century, James Clerk Maxwell deciphered the mathematical laws governing light's behavior. By the century's end, Peirce's work showed how it was possible experimentally to connect light with a measurement unit. The outcome, eight decades later in 1960, would be a new definition of the meter.

Events were set in motion by Michelson and Morley's improvements, news of which reached the BIPM shortly after the first formal meeting, in 1889, of its governing board, the General Conference on Weights and Measures (CGPM). Benjamin Gould, then the American member of the BIPM, visited Michelson's laboratory at Clark University and spoke with him about the possibility of Michelson's visiting the BIPM to continue his work. Gould then contacted the BIPM bureau's director René Benoît, who extended an official invitation to Michelson. Michelson arrived in 1892.

THE SPECTRAL LINE AND THE METER STICK

One of the actions of the CGPM at its first meeting was to formally accept the new artifact standards—The International Prototype of the Meter, and the International Prototype of the Kilogram—that had been manufactured just a few years earlier. These replaced the Meter and Kilogram of the Archives, which had been the formal standards since 1799. Another of the CGPM's acts was to distribute prototype standards to all of the countries that had signed the Treaty of the Meter. The first CGPM also initiated research into still better standards.

The leader of this effort was the Swiss scientist Charles Édouard Guillaume, who in 1883 came to the BIPM as an assistant to work on thermometer calibrations. In 1891, he began tackling the problem of developing better alloys for standards. In 1896, the chance observation of the special coefficient of expansion of an unusual bar of steel sent to the lab inspired him to mount a systematic investigation of alloys for possible use as standards. The result was his invention of a nickel-iron alloy which he called Invar; its low coefficient of expansion made it an excellent rigid material for the construction of standards and heavy-duty engineering equipment.

Guillaume staged several demonstrations of Invar's properties, including one, in 1912, which revealed the tiny vertical movements of the Eiffel Tower due to thermal expansion. He attached one end of a wire made of Invar to a support on the ground and another to the lever on the second platform of the tower, which was attached to a recording instrument. The sensitive equipment detected the effect not only of slight gusts of wind, but also of the tower's slight expansion and contraction due to temperature changes of just a few degrees. "Thus the Eiffel Tower appears as a gigantic thermometer of high sensitivity, notwithstanding its enormous mass," Guillaume later remarked.[1] From 1915 to 1936 Guillaume was director of the BIPM and in 1920 he won the Nobel Prize "in recognition of the service he has made to precision measurements by his discovery of anomalies in nickel steel alloys."

Meanwhile, Michelson arrived in Sèvres in summer 1892. His interferometer was damaged in transit, and his first task was to manufacture another. When he recommenced his measurements, he found the sodium line to be a composite of two different lines, which in his sensitive interferometer created fuzzy fringes, making it unsuitable for measurements of the degree of precision he was seeking. Michelson then sought another line that might be sharper. He tried a green line of mercury but this, too, produced fuzzy fringes. Finally, he settled on a red line of cadmium. Over the next year, he measured the red line of cadmium—a much sharper spectral line than the yellow sodium line Peirce had worked with—to 1 part in 10 million, and found that there were 1,553,164 such red cadmium lines in a meter.

This measurement made a strong impression on the Bureau's scientists. Participants at the second CGPM, which met in 1895, were delighted with Michelson's work and the possibilities it presented. They agreed that the Bureau should consider wavelengths of light as "natural representations" of the meter prototype. Participants at subsequent CGPM meetings frequently mentioned the desirability of connecting the meter with a natural standard. While much of the work of the BIPM in the first decades of the twentieth century consisted of carrying out calibrations of national standards, an increasing amount of research was devoted to developing the instruments needed for the *mise en pratique*—the "putting into practice," or the achievement of sufficiently reliable technology—that would be required for a definition of the meter in terms of light. The French scientists Charles Fabry and Alfred Pérot introduced several improvements into Michelson's interferometer and in 1906 remeasured the line of cadmium with a precision close to that available with the existing artifact standard.

The cadmium line was so sharp and reliably produced that for several years scientists at the BIPM and elsewhere focused their attention on it as the likeliest candidate for an eventual natural length standard: it became the spectral line of choice as a length standard, though the official length standard remained the meter in the vault. But did other elements produce superior lines?

In 1921, further improvements to the Fabry-Pérot interferometer allowed the BIPM scientist Albert Pérard to begin a systematic comparison and evaluation of potential spectral lines, including cadmium, mercury, helium, neon, krypton, zinc, and thallium. The results were surprising. Initially, scientists had assumed that—sodium's doubled line excepted—most spectral lines were equally sharp. They are formed by light emitted when electrons of an element jump from one specific energy level to another in the nucleus of an element; the wavelength of the light emitted is due entirely to the difference between the energy levels. ("I'm not a very funny guy," runs an old piece of wit among spectroscopists, "I only know a few good lines.") If so, the only reason to prefer one spectral line over another would be the ease with which it could be produced and detected.

Pérard and others found this not to be the case. Naturally occurring elements consist of a range of isotopes, which have the same number of protons but different numbers of neutrons, and the difference between the nuclear structures of isotopes of the same element means that the spectral lines are blurred, with slightly different energies. Other magnetic features of the nucleus—called hyperfine structures—also affect the energy levels of states and blur spectral lines, though hyperfine structures were expected to be minimal in nuclei with even atomic numbers. Scientists found a further source of line blurring in the Doppler effect: atoms are always in motion, always vibrating—and as the atoms producing the light approach and recede from the light-gathering instruments, their wavelength appears to shorten and lengthen. The lighter the element, the more it moves and the greater the Doppler effect and the resulting broadening of its spectral lines. Light was proving more complicated than expected.

These findings sent metrologists hunting for a heavy element with few and rare isotopes and even atomic numbers. Throughout the 1920s and 1930s, metrologists in three countries examined several different spectral lines for a possible redefinition of the meter: in the United States, researchers at the NBS were examining mercury 198; in Germany, they were examining krypton 84 and 86; and at the BIPM, cadmium 114. Metrologists expected that discussion of these candidates, and even a possible decision, would take place at the ninth CGPM, which should have taken place in 1939 but was cancelled with the outbreak of World War II. When the ninth CGPM finally took place in 1948, its participants had much to tell each other, including the unexpected news that a recent measurement of the Kilogram of the Archives revealed it to have lost weight, evidently due to the escape of bubbles trapped in the platinum. The Conference, whose work had been disrupted by the ravages of World War II, was not ready to fix an ultimate standard of length; its scientists were not convinced that the precision available—about a part per million—was consistent and reliable enough in the technology distributed throughout the world to make it a rival of the artifact prototype. The ninth CGPM therefore recommended further work on this technology, and requested that national metrological laboratories continue to study the instrumentation to produce and measure the spectral lines, and the spectral lines themselves.

By 1952, enough additional work had been completed that BIPM scientists created an advisory panel to plan the eventual redefinition of the meter. The panel was named a "Consultative Committee," one of several that the BIPM had established: for electricity in 1927, photometry in 1933, and thermometry in 1937. Participants at the next CGPM, the tenth which took place in 1954, reached the consensus that it would be possible to formally redefine the meter in terms of a natural standard at the following CGPM, the eleventh, scheduled for 1960.

In his book *Philosophical Investigations*, the Austrian philosopher Ludwig Wittgenstein wrote that "There is one thing of which one can say neither that it is one meter long, nor that it is not one meter long, and that is the standard meter in Paris" (Section 50).[2] He was not making a point about the meter stick itself—that it somehow has the extraordinary property of being without a length—but about the practice of measuring: while we are in the process of using a standard to measure the length of something, it does not make any sense to attribute a length to what is used as the standard. The book's appearance in 1953, two years after Wittgenstein's death, came uncannily at the same time as plans for permanently retiring the meter stick were being formulated. The philosophical point would now be transferred to the wavelength of light. After the redefinition, planned for 1960, would take place, the wavelegth of the chosen spectral line would no longer be measurable; it would be the ruler, not the ruled.

IMPEDIMENT TO IMPERATIVE

The 1950s were also a time in which many nations worldwide were converting, or preparing to convert, to the metric system. Initially, the metric system's "all-conquering march," Kula wrote, was imposed by force; "it marched in the wake of French bayonets."[3] Other nations eventually began adopting it out of more positive motives, such as fostering national unity, repudiating colonialism, enhancing international competitiveness, and as a necessary precondition for entering the world community. By the middle of the twentieth century, the metric system as envisioned by the French revolutionaries was indeed becoming universal, en route to being adopted by nations all over the globe. Our two contrasting examples—West Africa and China—continue to serve as a synecdoche for the conversion of dozens of other nations to the metric system.

Most of Africa would convert to the metric system in the early 1960s. The emerging independent African countries saw it simply as a precondition for throwing off colonialism and entering the international community. Many West African lands converted at the time of their independence in 1960; Ghana was one of the few holdouts and did not fully convert until 1975.

China's route to the metric system was far more complicated. The Opium Wars left China weak and debilitated, and the Qing dynasty, whose legitimacy was already shaky, suffered another blow when it lost the first Sino-Japanese war of 1894–95. In 1898, a reform movement led by a young emperor named Guangzu was crushed by conservative opponents led by the Empress Dowager Cixi. Struggling to hold the Qing dynasty together, she introduced a series of reforms which included metrological reforms. She ordered China's ambassador to Paris to visit the BIPM to seek advice on conversion to the metric system and asked for two pairs of rulers and weights for the dynasty. Meanwhile, in 1908, the Qing dynasty redrew its laws in a way that reorganized the country's system of weights and measures. The dynasty retained the traditional Chinese weights and measures, but defined them in terms of the metric system, stipulating the ratios between the traditional Chinese metrological units and those of the metric system.

In 1909, the new standards arrived in China, just in time for the end of the Qing dynasty. For in 1911, the Wuchang uprising, prompted to a large extent by popular outrage at the inability of the dynasty to set limits on the incursions of foreign powers into policy matters like weights and measures, led to the Xinhai Revolution, which in turn brought about the end of the Qing dynasty and the creation of the Republic of China in 1912, whose first provisional president was Sun Yat-sen. The new government was also concerned about the disunity of Chinese weights and measures and continued its predecessor's contacts with the BIPM, sending its own representatives to the BIPM in 1912. The government created a new agency, the Bureau of Measurement, to improve the country's weights and measures system.

Moving the metric system into the hearts and minds of the Chinese people proved vastly more difficult than the republicans hoped. "The problem was not resistance from the Chinese people," Zengjian Guan, a professor in the Department for the History and Philosophy of Science at Shanghai Jiao Tong University, told me. "The main reason why it took so long to accomplish the transition to the metric system was the social upheaval of China at the time, the continuous wars and revolutions that the country was experiencing."[4] Furthermore, China and other Asian nations, too, have occasionally had their own equivalent of pyramidologists, who claimed that ancient Asian scientists had discovered the fundamentals of science well before contact with the West; some even declared that these principles, discovered first in Asia, had been communicated to the barbarians in the West in the distant past.

In 1925, Sun Yat-sen died, and Chiang Kai-shek took his place. Chiang began ruling China with an iron fist, and in 1927 established a new government seat in Nanjing. Chiang's government, too, placed a high priority on unifying weights and measures, and in 1929, issued a law that kept the traditional Chinese measures in place for internal use, but put the metric system into play for official transactions. The second Sino-Japanese war, which began in 1937 and which had displaced Guangming's family as well as thousands of other families, brought further attempts at conversion to a halt.

After World War II, the People's Liberation Army led by Mao Zedong fought Chiang's government and at the end of 1949 succeeded in driving it entirely out of mainland China. The ensuing People's Republic of China, too, was interested in unifying the country's weights and measures system, and in fully converting to the metric system, and embarked on an attempt whose initial phase was completed by 1959. "It's true that those years were a time of antipathy to things Western," Zengjian told me, "but that mainly involved matters of politics and lifestyle. The PRC was very interested in developing science and technology, and did not care whether this came from the West or any other place. By then the metric system was central to world science and technology, so its origin did not pose a problem to its introduction into China." Still, political turmoil lasting through the Cultural Revolution of the 1970s impeded full conversion, which had to await passage of the Act of Measurement in 1985.

I asked Guangming, who lived through these attempts, if the transition was easy. "No," she said. "But once again, the Chinese were clever. The leaders told the people that the new metric measures were just like the old Chinese measures according to the 1, 2, 3 system: 1 sheng of volume was a liter, 2 jins were a kilo, and 3 chis were a meter. When I was a child, I remember thinking, 'How smart we Chinese are, to have had such accurate measures in ancient times—we were much smarter than the Americans!' Of course it wasn't true; the leaders fudged the old system. But it made it easy to convert."[5]

Amazingly, the United States—a worldwide leader in science and technology—was one of few holdouts. The reason was familiar; the fact that it was a leader meant that politicians viewed conversion with little urgency. Nothing seemed broken; what was to fix? U.S. scientists all used the metric system, and in normal commerce it was easy enough to convert in those few cases where it was necesaary. Furthermore, American politicians tend to abhor reforms that cost money. Nevertheless, in the 1950s the time again seemed ripe. During the cold war, politicians viewed maintaining technological superiority as essential to military defense—and scientists considered use of the metric system an indispensable element to a modern approach to technology.

The launch of the Soviet spacecraft *Sputnik* on October 4, 1957, generated fears of gaps between American and Soviet technology. Senator John F. Kennedy appealed to (unjustified, it would turn out) fears of a "missile gap" in his Senate campaign of 1958 and presidential campaign of 1960. The Soviets were thought to have more intercontinental ballistic missiles, with heavier warheads, than we had; they were also the first to launch rockets that impacted the moon and that took photographs of its far side.

Additionally, the Soviets were considered a threat to the free world due to a "measurement gap," which was possibly more dangerous than the missile gap. Missiles, after all, depend on precise measurement. During the previous decades, the demand for precision had shot up. At the end of World War I, it was rare for equipment to require tolerances much beyond about 1 part in 10,000. By the 1950s, high-technology equipment was beginning to require tolerances of 1 part in 100,000 or even 1,000,000 (a millionth of an inch is about what you would get if you took a human hair and divided its width into 3000 equal strands). The requirements of the space age upped this yet another notch, to 1 part in 10 million or even 1 part in 100 million. American scientists, educators, and businessmen were concerned that the U.S. failure to adopt the metric system would hamper its capacity for technological innovation, science education, and industrial competitiveness. In 1959, U.S. Secretary of Commerce Lewis Strauss announced his support for American conversion to the metric system, and legislation was introduced into Congress for a program of research into U.S. conversion. Meanwhile, the United States, Great Britain, and several other nations still using the imperial system agreed to revised standards for the imperial system that defined them in metric terms, with 1 inch equal to 2.54 centimeters and 1 pound equal to 0.45359237 kilograms.

In 1960, when Beverly Smith, the Washington editor of the *Saturday Evening Post*, asked an engineer why the Soviets were winning the missile race, the answer he received was that "we can't measure well enough" and that "the Russians may be ahead of us" in measurements. While the Soviets had a vast network of several measurement laboratories, our own National Bureau of Standards was underfunded and facing further cuts. Smith darkly warned that it was possible that "the future direction of civilization might turn on millionths of an inch or billionths of a second."[6]

Edward Teller (1908–2003) was an equally dire prophet. Teller was a Hungarian-born physicist who moved to the United States in 1935, who worked on the Manhattan Project, and whose way of acting patriotic about his adopted country was to warn that if the United States did not embark on certain drastic actions that he championed, it would abandon world leadership to the Soviet Union. These actions included development of the hydrogen bomb, extensive nuclear weapons testing, and promoting the flawed Strategic Defense Initiative to protect the United States from space. They also included the conversion to the metric system.

"This is a subject on which I am rabid," Teller declared. The Soviet Union had abandoned its "versts and other absurd measurements" for the metric system in 1927, with dire implications for the free world. Its use of the metric system, at a time when we are hobbled by our dependence on the imperial system, is a "Red weapon" that helps Soviet efforts to forge ahead of the United States in technology, education, and commerce. U.S. conversion is "urgent" and may spell the difference in "the fight for our ideals." At stake is nothing less than "the greatest contest in the history of the world—the contest for leadership of the world itself."[7]

Teller's championing of the metric system was a rare instance in which he could not bully nor muster significant support among his usual political allies. U.S. resistance to the metric system defeated even Teller. Though measures for U.S. metric conversion were periodically introduced into Congress, year after year passed with no action.

THE SI

The fears of a "measurement gap" sparking global conflict between the United States and the Soviet Union could not have been in sharper contrast with the international and collaborative atmosphere taking place at the BIPM during those same years. The organization had been established to take national rivalries out of measurement, and it had succeeded brilliantly in that mission. The group tended to act slowly, cautiously, and only when consensus had been reached. From the beginning, in 1875, its delegates had collaborated peacefully even when their respective governments were belligerent, and even officially at war. During World War I, when Germany and France were fighting, one of the three keys to the safe containing the international standards was in the hands of a German representative. When Paris came under German artillery fire, and Bureau officials considered but rejected moving the standards and copies to a safer location, they had an extra set of keys made in case such a move ever became suddenly necessary in wartime.

At first, the Bureau's business was essentially taking care of the meter and kilogram prototypes, comparing these standards with those of member states, and developing measures of volume, density, and temperature. But as a neutral, international agency, it became the organization other countries turned to for any questions about other measurement issues. In the early twentieth century, it was receiving requests to extend its scope to cover other kinds of units with international implications.

By the 1920s, for instance, electrification and demands placed on the growing electrical industry created the need for standardization of electrical units. Several different systems of electrical measurements were in use, and the BIPM was the logical institution to act as adjudicator. In 1921, participants at the sixth CGPM amended the Treaty of the Meter to give the Bureau authority to establish and conserve prototype standards of electrical units and their copies, and carry out comparisons of these with national standards. This considerably enlarged the Bureau's mission. The CGPM also expanded the metric system to incorporate the second and ampere in a general framework called the MKSA—meter, kilogram, second, ampere—system.

As science and technology evolved further, the Bureau faced requests to standardize other types of measurement, including time, light intensity, temperature, and ionizing radiation. Often its work consisted of ratifying agreements that had been made between national metrological laboratories. By the ninth CGPM meeting in 1948, it was clearly desirable to integrate all these measurements into a comprehensive system of units, which in turn could form the basis for new derivative units for purposes yet unknown. The tenth CGPM in 1954 took important steps toward that end by adopting three base units in addition to the meter for length and the kilogram for weight: the ampere, degree Kelvin, and candela. The ampere was formally defined as "that constant current which, if maintained in two straight parallel conductors of infinite length, of negligible circular cross section, and placed 1 meter apart in vacuum, would produce between these conductors a force equal to 2×10^{-7} [newton] per meter of length." The degree Kelvin (soon simply *kelvin*), a unit of thermodynamic temperature, was defined such that "the fraction 1/273.16 of the thermodynamic temperature of the triple point of water." The candela was defined such that "the brightness of the full radiator at the temperature of solidification of platinum is 60 candelas per square centimeter."

On October 14, 1960, the thirty-two delegates to the eleventh General Conference, deciding that the current meter was not defined "with sufficient precision for the needs of today's metrology" and that it was "desirable to adopt a standard that is natural and indestructible," ratified the following resolution: "The metre is the length equal to 1650763.73 wavelengths in vacuum of the radiation corresponding to the transition between the levels $2p_{10}$ and $5d_5$ of the krypton-86 atom." At last, the meter was tied to a natural standard. (The meter was redefined once again in 1983, as "the length of the path traveled by light in vacuum during a time interval of 1/299,792,458 of a second.") The platinum-iridium bar, the international prototype meter, which had reigned over the international network of length measures since 1889, became a historic object; the new standard was universal, everywhere, not localized. What was localized was the technology needed to produce it; the assumption was that this technology quickly would become available worldwide.

This was a key decision in the history of metrology. It was the culmination of centuries of thought and technological development: from the first ideas expressed about the desirability of connecting measurement units to natural standards in the seventeenth century, to the attempts made in the eighteenth, to the realization of the impossibility of this dream given the technology in the nineteenth century, to its rebirth and final realization in the twentieth.

Press coverage of the CGPM decision was widespread. Many reports quoted a remark by an American delegate who attempted to communicate the importance of the change by saying that the smallest measurement that could be made with the platinum-iridium bar standard was a millionth of an inch, and "an error of one-millionth of an inch in the borehole of a guidance gyroscope could cause a space shot to miss the moon by a thousand miles."[8] The *Chicago Daily Tribune* complained, "We get the feeling that important matters are being taken out of the hands, and even the comprehension, of the average citizen." Woe to the color-blind seamstress, it continued, who can use a tape measure but can't tell an orange-red wavelength.[9] The half-joke concealed the worry that measurement matters, which should be simple for the average person to understand—that had been, after all, one of the principal motives for constructing the metric system—were about to become too complex for anyone except scientists.

While the adoption of the first natural standard attracted widespread hoopla, seeming to be the big step for metrology at the time, the second step of the CGPM was in a way much more radical and significant—the reorganization of units. In essence, the CGPM replaced the metric system with a new, correlated system of units that together provided the framework for the entire field of metrology, mechanical and electromagnetic. The system consisted of six basic units—the meter, kilogram, second, ampere, degree Kelvin, and candela (a seventh, the mole, was added in 1972.)[10] It also included a set of derived units built from these six, with their own special names.

Before 1960, the second had been defined in astronomical terms, as 1/86,400 part of the day; now it was 31,556,925,974 part of the solar year 1900. But the technology for measuring time was rapidly advancing,

SI BASE UNITS AND PREFIXES

UNITS	PREFIXES			
	yotta (Y)	1,000,000,000,000,000,000,000,000	(added 1991)	10^{24}
	zetta (Z)	1,000,000,000,000,000,000,000	(added 1991)	10^{21}
	exa (E)	1,000,000,000,000,000,000	(added 1975)	10^{18}
	peta (P)	1,000,000,000,000,000	(added 1975)	10^{15}
	tera (T)	1,000,000,000,000		10^{12}
	giga (G)	1,000,000,000		10^{9}
	mega (M)	1,000,000		10^{6}
	kilo (k)	1,000		10^{3}
	hecto (h)	100		10^{2}
	deca (da)	10		10^{1}
meter (m, length)				
kilogram (kg, mass)				
second (s, time)				
ampere (A, electric current)		1		
kelvin (K, thermodynamic temperature)				
candela (cd, luminous intensity)				
mole (mol, amount of substance)				
	deci (d)	0.1		10^{-1}
	centi (c)	0.01		10^{-2}
	milli (m)	0.001		10^{-3}
	micro (μ)	0.0001		10^{-6}
	nano (n)	0.000,000,001		10^{-9}
	pico (p)	0.000,000,000,001		10^{-12}
	femto (f) (added 1964)	0.000,000,000,000,001		10^{-15}
	atto (a) (added 1964)	0.000,000,000,000,000,001		10^{-18}
	zepto (z) (added 1991)	0.000,000,000,000,000,000,001		10^{-21}
	yocto (y) (added 1991)	0.000,000,000,000,000,000,000,001		10^{-24}

The seven current SI base units and their prefixes. The kilogram, again, is idiosyncratic in being a base unit and still having a prefix. The SI also includes units derived from base units through multiplication and division. These include units for area (square meters or m^2), volume (cubic meter or m^3), velocity (meter per second or m/s, acceleration (meter per second squared or ms^2) and so forth. The full list can be found on the BIPM Web site at http://www.bipm.org/en/si.

with the first atomic clock built at the NPL, Great Britain's national metrology laboratory, in 1955. The thirteenth CGPM, in 1967, would redefine the second yet again: "the second is the duration of 9 192 631 770 periods of the radiation corresponding to the transition between the two hyperfine levels of the ground state of the cesium 133 atom." Time therefore became the next measure to be tied directly to a natural standard.

The eleventh CGPM in 1960 faced the question of what to call this new reorganization and extension of measures. The name "metric system" had referred to the units for length and mass. What the CGPM had created was much more comprehensive, and after some discussion this new system was called "The International System of Units," or SI after its French initials. For the first time, the world had not merely universal units but a universal system of units.

CHAPTER ELEVEN

THE MODERN METROSCAPE

The story of measurement encompasses more than the tale of how today's network of standards, instruments, and institutions came into being. It also includes the changes that take place in the *meaning* of measurement. Every age has a metrosophy, a shared cultural understanding of why we measure and what we get from measuring, and the understanding evolves over time.

But these shared cultural understandings are more difficult to talk about, especially since each age is convinced that it doesn't have any—that it has evolved beyond metrosophy. "The way *we* measure is the right way, and connects us with reality," we tell ourselves. Even Witold Kula, that eminent and sensitive decipherer of the social logic of European medieval measures and how closely they are woven into human life, shared this view. True, in *Measures and Men* he often drops remarks to the effect that the metric system is "sheer convention," has "no practical social meaning," lacks a connection to "social values," amounts to "dehumanization," and has "no inherent social significance whatsoever." Nevertheless, he grudgingly admits to being an "admirer" of the metric system, which has brought about a "higher level of mutual understanding among people," and "taken us very far along the road of more effective and fruitful international understanding and cooperation."[1] The book's final sentence—"And in the end, a time will come when we shall all understand one another so well, so perfectly, that we shall have nothing further to say to one another"[2]—however ironic, makes it clear that Kula faults the modern world, not the modern system of measurement so perfectly adapted to its realities.

But the great scholar has nodded. The modern system of measurement is not devoid of social meaning, of metrosophy. The thoroughgoing project of stripping the imprint of regions, products, and times from measures, of abstracting measures from each and every local context in order to make the world measurable, calculable, and universal for human beings and to put it at our disposal, has a deep social meaning indeed.

THE NEW VITRUVIAN MAN

Henry Dreyfuss's (1904–1972) "models" help bring this hidden social meaning to light. Dreyfuss was a no-nonsense, no-frills industrial designer of the mid-twentieth century whose well-received practical products included the Hoover vacuum cleaner, the John Deere tractor, and the Princess telephone. His distinctive approach was to investigate human measurements and use them to orient the design of appliances and equipment. His books, which include *Designing for People* and *The Measure of Man*, provide line drawings of a pair of archetypical human beings, named "Joe" and "Josephine," whose accompanying measurements (the fruit of decades of data collection and research) were intended to allow engineers to incorporate human form and behaviors from the start into products and machinery. "They are not very romantic-looking, staring coldly at the world, with figures and measurements buzzing around them like flies," Dreyfuss writes, "but they are very

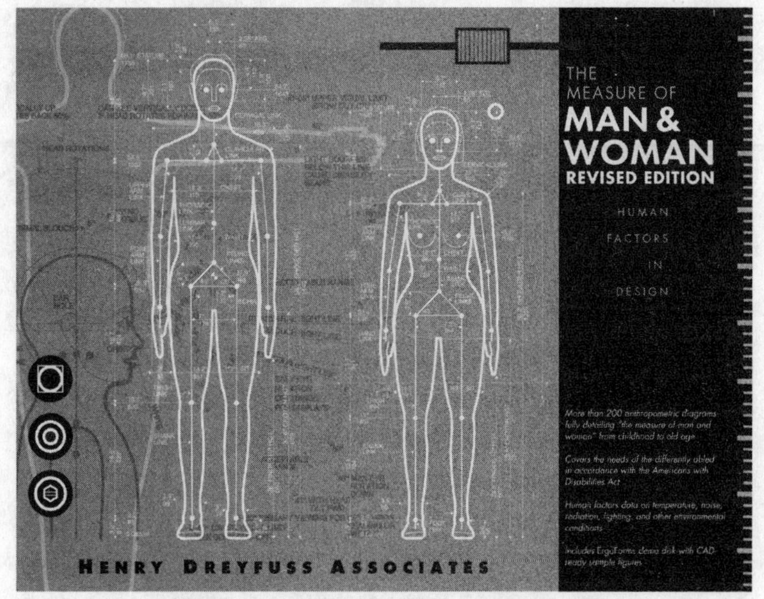

"JOE" AND "JOSEPHINE," WHOSE MEASUREMENTS ARE USED TO DESIGN BETTER HUMAN-MACHINE INTERFACES.

dear to us." Dear, indeed, to engineers who must figure out the optimal design for phones and irons, tractors and aircraft. "[T]he most efficient machine," Dreyfuss continues, "is the one that is built around a person." Editions of *The Measure of Man*, published after the designer's death and renamed *The Measure of Man and Woman*, include measurement data for young and old humans, for the handicapped and impaired, and for the "percentile person" whose dimensions are in extreme (1 to 99) percentiles. A CD included with the book contains sample CAD-ready figures to be cut, pasted, and adapted. Still more effective are three-dimensional body scans of Joe and Josephine, a technology created after Dreyfuss's death, that has already advanced dramatically. Joe and Josephine's dimensions are currently available online and are continually updated.

Joe and Josephine are the new Vitruvian Man, gendered into a couple and measured over their entire lifespans. The metrosophy of their world is much different from that of the ancient Greeks. Joe and Josephine are utterly devoid of the boldness, nobility, and beauty of Leonardo's Vitruvian Man. Their reason for being—why they were created and what they show us about the world—is not beauty and symmetry but efficiency. They do not help connect us humans with something beyond the world, but help engineers and the individual human beings for whom the engineers design to get a better grip on this one.

Dreyfuss's career unfolded during several decades of great enthusiasm for the use of human measurements to increase efficiency. Frank Gilbreth (1868–1924) was another pioneer. An industrial psychologist and efficiency proponent, Gilbreth created a special unit called the Therblig, referring to standardized human body movements and named after himself (it's the backward spelling of his last name, more or less). Here's a passage from the novel *Cheaper by the Dozen*, based on Gilbreth's life:

> Suppose a man goes into the bathroom to shave. We'll assume that his face is all lathered and he is ready to pick up his razor. He knows where the razor is, but first he must locate it with his eye. That is "search," the first Therblig. His eye finds it and comes to rest—that's "find," the second Therblig. Third comes "select," the process of sliding the razor prior to the fourth Therblig, "grasp." Fifth is "transport loaded," bringing the razor up to the face, and sixth is "position," getting the razor set on the face. There are eleven other Therbligs—the last one is "think"! When Dad made a motion study, he broke down each operation into a Therblig, and then tried to reduce the time taken to perform each Therblig. Perhaps certain parts to be assembled could be painted red and others green, so as to reduce the time required for "search" and "find." Perhaps the parts could be moved closer to the object being assembled, so as to reduce the time required for "transport loaded." Every Ther-

blig had its own symbol, and once they were painted on the wall Dad had us apply them to our household chores—bedmaking, dishwashing, sweeping, and dusting.[3]

The new Vitruvian couple do not live in a world where measurement simply describes the beautiful relationships of their bodies. Measurement creates the world they live in, its devices and environment, how they act in it, and the way they understand it. Measurement therefore structures their answers to the famous three key questions that, the philosopher Immanuel Kant tells us, occupy all human reason: "What can I know?" "What must I do?" and "What may I hope?" Measurement is not merely one tool among others, belonging only to separate elements such as rulers, scales, and other instruments; measurement is a fluid and correlated network that is smoothly and intimately integrated into the world and its shape.

The modern world, in short, has a *metroscape*. The suffix "-scape" commonly refers to a kind of space (think landscape, seascape, or cityscape) that is extended, produced by civilization—human interaction with nature—has a particular character, and shapes how human beings relate to nature and each other. In words like *soundscape* and *ethnoscape*, the suffix has also been applied to more virtual kinds of spaces with a similar impact on human life. A "-scape" is neither simply material nor mental but both at once; it inhabits the world and its features and simultaneously the way we perceive and relate to this world. The modern metroscape is not the doing of the SI, which is a consequence rather than a cause of the metroscape.

In Chapter 4, I mentioned that the availability of standard measures and a network of institutions to govern them was essential to the division of labor in the new economic and political milieu of Europe in the eighteenth and nineteenth centuries. What was true of Adam Smith's famous pin factory was true of myriad other workplaces. The measurement system shaped products, workers, markets, and businesses, and both reflected and reinforced social, political, and economic forces. The metroscape, in short, was key to the emergence of capitalism. It is vital, too, in contemporary agriculture. Lawrence Busch, a sociology professor at Michigan State University and director of the Center for the Study of Standards and Society, and Keiko Tanaka have described the role of measurements and standards in the agriculture of canola, a seed used to produce oil. "[G]rain grades link farmers to elevator operators. Seed quality tests link seed producers to farmers. Measures of oil content and composition link large sellers to buyers of canola. Measures of shelf life link processors to retailers. . . . Tests are measures of nature at the same time as they are measures of culture."[4] The kind of role for measures that they detect in the production and consumption of canola is found in nearly every other agricultural product.

CLOTHES

The metroscape appears in clothes manufacture also. To take just one example, the use of 3-D scanners prophesied by earlier editions of *The Measure of Man and Woman* has become more practical, and marketable, at the beginning of the twenty-first century, and has begun to change how clothes are made and produced. The technology is the result in part of a government-sponsored basic research program to support troubled national industries. In 1979, a National Science Foundation–sponsored study of the textile industry found that it scarcely did any R&D, raising concerns about global competitiveness. Two years later, a nonprofit R&D organization was created to help: the Tailored Clothing Technology Corporation or $[TC]^2$, funded jointly by government and industry. Its first major quest was to deflect the flow of apparel-producing jobs to low-wage countries by developing equipment to robotically manufacture men's suits. That attempt ultimately failed (as did a similar one in Japan) to produce automated suits or to deflect the jobs flow, though [TC]^2 did manage to automate production of things like sweat pants and T-shirt sleeves.

But [TC]^2 kept up its research program, and by the end of the 1990s, its 3-D body technology could be used in retail stores. The first machine was installed in a Levi Strauss store in San Francisco in 1999. Buying jeans went high-tech.

Two years later, Brooks Brothers Madison Avenue installed the first 3-D scanner in New York City. The tradition-bound company was then struggling to modernize and cut expenses without compromising quality. Senior staff were uneasy with the technology at first, according to Joe Dixon, senior vice president of Production and Manufacturing, who initially operated the device himself. Part of their reluctance sprang from fear that the scanner would undermine the company's stance: a firm midpoint between bespoke (custom-made) and made-to-measure clothing, famously characterized by New York designer Craig Robinson as that between "tradition and personality versus conformity and convention."[5]

Dixon found that the new 3-D measuring technology had little to do with this distinction, which was always a spectrum rather than an opposition; the scanner's impact was merely to shorten the time-consuming fitting process for busy New York professionals. Women, too, dropped by, even though the machine wasn't intended for them. "I got propositioned," Dixon told me. "Several women said something to this effect: 'If you can make me a pair of pants that fit me perfectly, I'll be yours forever!' That impressed me with just how particular people can be about how their clothes fit."[6]

Institutions with no political or emotional commitment to the bespoke versus made-to-measure distinction embraced the machine. The Coast Guard induction center at Cape May, New Jersey, purchased two—one for men, one for women—so recruits can be lined up in their underwear for faster scanning. More than forty universities now have scanners, mostly in apparel or fashion departments.

The scanners are expensive; the first Brooks Brothers machine cost $75,000 and took up 140 square feet. This delayed their use in mainstream stores, and for 8 years Brooks Brothers had the only one in New York. Still, they did enough business that the company upgraded to a newer model, which costs $30,000 and takes up 20 square feet, and outfitted ten of its stores with space for body scanners. The drop in cost and valuable store space increased demand. Scanners built by [TC]^{2}, the U.S. market leader (the world market leader is a German company called Human Solutions), occupy over 100 locations worldwide.

After Brooks Brothers, the next two scanners to appear in Manhattan were at Alton Lane and Victoria's Secret. Alton Lane is a small menswear company that aggressively uses the scanner to promote a bespoke approach, trying to reach a high-volume market by doing for custom suits what Netflix did for movies and Blue Nile (an online retailer of expensive jewelry) for diamonds. It exploits the "Beam me up, Scotty" technological experience as an attraction, creating Web accounts for its customers. Customers can order a traditional customized suit or create their own at home, using their 3-D image to "try on" different kinds of lapel, trouser break, cuffs, and so forth, with the image updating live, manipulating it to see what the suit looks like on them from all angles. Alton Lane hosts events where partygoers are individually scanned as the warm-up activity, followed by wine and cheese. Victoria's Secret in SoHo installed a scanner at about the same time as Alton Lane, with the intent of helping women to find the perfect-fitting bra.

My wife and I decided to have ourselves scanned, and compare notes.

We often experience the same thing differently, but few as strikingly as what happened in our body scan experiences.

Mine took place in the Brooks Brothers flagship store on Madison Avenue. I undressed in a fitting room, donned a specially colored undergarment called scanwear, entered a dark booth, and grasped handles that fixed my position. When I pressed a button, patterns of light from sixteen different sensors played around my body for almost a minute, producing between 600,000 and 700,000 data points accurate to two-tenths of a millimeter. By the time I had dressed and emerged, a computer had smoothed, filtered, and compressed the data into a 3-D body image; I got a printout, along with lists of measurements. These were only numbers, but they were mine, and I could use them to virtually try on and order suits and other clothes made from scratch to fit my particular shape. The experience was Disney-like: comfortable, smooth, dazzling, and leaving me feeling entertained and even special.

Down in SoHo, my wife stepped from a plush pink fitting room into a black closet, and as the lights flashed a recorded voice assured her that the scan's measurements would result in the best-fitting bra she ever had; "Body Match" is the company's name for the scanning process. She was then handed a card listing six off-the-shelf items, which the salesgirl dutifully brought to her. Two fit, one fabulously.

Same technology, different experiences. My wife and I wound up at the opposite poles of the bespoke and made-to-measure divide, and in terms of customer appreciation. Her experience lacked the luxe appeal of a Parisian lingerie salon or the sensitive hands of an experienced fitter; it was less Disney than airport or doctor's office. Sometimes, customer experience has to be integrated into a technology for it to succeed. Otherwise, the numbers just don't add up.

Not long after, I witnessed a showdown between the two most recent 3-D scanners of [TC]² and Human Solutions. It took place in the industrial town of Haverhill, Massachusetts, home of Southwick, the Brooks Brothers manufacturing facility. Joseph Antista, Southwick's director of training, had asked the two companies to set up their latest models in the main office.

As I walked in the office door, the [TC]² model stood to the left, a 7.5-foot-tall, black, closetlike booth about 4 by 5 feet. To the right was its challenger, behind an olive-green curtain that cloaked three 9-foot towers positioned in a triangle about 7 feet on a side. Each model was designed by a team that included one physicist, computer programmers, and several engineers. Each requires about an hour to set up, but could be plugged into a standard electrical outlet. Each needs but a few seconds to scan a customer.

Fitting is a dying art, Antista told me, and body measuring is time-consuming and expensive, part of the custom-clothing process that seems amenable to technologization. He was therefore testing the two devices (Cyberware, a distant competitor based in Hollywood, caters to a more high-tech niche) for possible use in Brooks Brothers' 100 or so stores and numerous outlets.

While he studied data, I ran my own independent test.

I tried [TC]² first, whose technology is "Structured Light Assisted Stereo 3-D Sensing." I undressed, put on the scanwear, and stepped into the booth. I could make out sixteen sensors embedded in the walls in front of and behind me. Each sensor had two cameras which looked at me from a different angle.

A pleasant recorded female voice welcomed me and instructed me to stand in place, grip the handholds, and push a button. Lights flashed, projecting patterns on my body while the camera gains were optimized for my skin color. Then the image acquisition process began, with striped patterns used to "structure" the imaging to facilitate triangulation. While stereo sensing with white light is a well-known range-

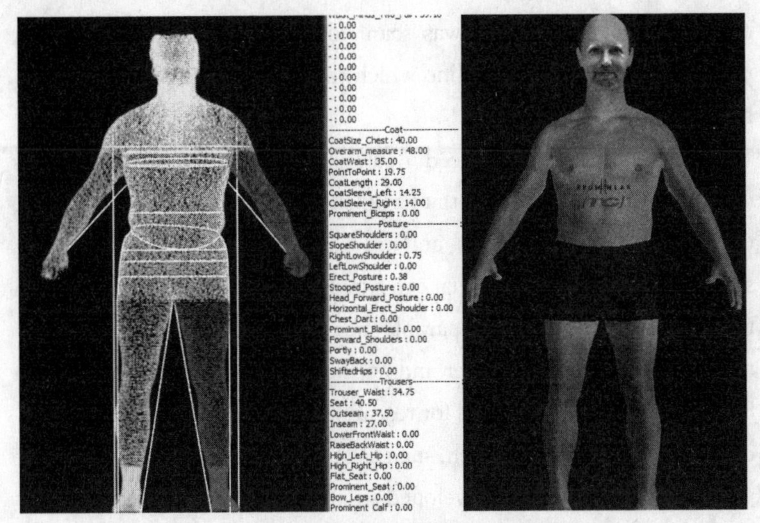

THE AUTHOR'S DATA SET AND AVATAR AS RECORDED,
WITH PITILESS PRECISION, BY A [TC]² 3-D BODY SCANNER.

finding technique—it was used, for instance, on the Mars Rover—it is computationally intensive, and only recently became fast enough to use in body scanning.

The experience, which took about a minute in all, was "quick, easy, safe, private," as the [TC]² vice president for technology development David Bruner assured me it would be.[7] The device has no moving parts. To use white light, Bruner explained, the technology was highly optimized for scanning human beings in upright or seated poses; it would not work well for, say, animals.

Then I tried the Human Solutions device, which uses lasers rather than white light. Roy Wang, a Human Solutions representative who majored in physics at the University of Toronto, explained that triangulation is easier with lasers—though some markets, such as the U.S. one, unreasonably fear lasers for alleged (though unfounded) safety reasons. Each of the three towers in the device has a laser and camera mounted on a moveable sled. As I was scanned, the sleds slowly descended, projecting a horizontal red line which the cameras used to acquire my body's cross-section.

Both technologies produced a data cloud of hundreds of thousands of points accurate to one-tenth of a pixel, sorted them into a 3-D model, extracted key measurements, and produced an "avatar"—an aesthetic, skin-covered visualization that can simulate how clothes would look on it. Both machines have comparable measurement accuracy. The [TC]^2 machine is somewhat cheaper and optimized for clothing. The Human Solutions model is optimized for research purposes and has found applications in ergonomic research, such as at NASA's Houston facility.

Human Solutions has developed simulations involving moving avatars. "Women, being curvier than men, may want to see how belts and appliqués will look in different positions while in motion," Wang said. "It's not trivial!" To illustrate, he put a belt and an appliqué on a simulated female avatar, and varied the properties of her clothing, such as the fabric's elongation, compression, and bending rigidity. The look changed dramatically. "Now watch this!" He tweaked the dials so that the simulated woman's clothes snagged on a tree branch, to test how the fabric reacted.

The one negative moment of my experience was seeing a pitilessly precise 3-D model of my near-naked body. My reaction was not atypical, Antista said. "The technology's *too* good," he said as he showed me around the floor of the factory. "It shows you as you are. Whichever scanner we decide to use, we'll be dressing the client's avatar in their chosen clothing from the start."

I asked him if the measurements were in any way inferior to those of fitters. "Their skill was not in the measuring," he replied. "It was in knowing what would happen when you put the clothes on a person with a particular posture and character. A measurement helps a fitter decide,

'What am I going to put on her?' A person is not a piece of wood. A person is alive."

Antista stopped by a seamstress and picked up part of a suit she was preparing. "See the fabric? It's alive, too. It has seven layers of different fabric that reacts differently when it moves, and changes when it's cleaned. A suit is also not a piece of wood."

Body scanning, yet another example of an area in which the availability of massive, inexpensive computing and measuring power has made heretofore high-tech applications of physics more routine, is quickly growing past custom clothes manufacture into marketing. Its use is going hand-in-hand with the shrinking of retail space and the ever greater connection of store and online offerings. The 3-D technology, in turn, is giving birth to a cross-industry adoption of body scanning, linking health, fitness, medical, entertainment, and gaming institutions with apparel, in a new area of the metroscape.

LIVING STANDARDS

As my wife discovered, measurement is particularly tricky when it comes to brassieres. In the apparel industry—a vast economic and industrial complex—more resources go into testing and evaluating bras than any other product. The reason is that bra selection involves a nexus of aspects involving fashion, comfort, self-esteem, and self-image unlike any other article of clothing, making their measurement harder to standardize (a problem completely absent in the case of, say orchidometers, the devices used to measure testicle volume). In the past decade, engineers have applied to bras the 3-D technology used to plan astronaut spaceflights, building scanners that can track locations of forty separate data points on a moving breast, with sensors to detect skin pressure and deformation when wearing bras. This is the latest in a (so far futile) attempt to automate brassiere designing and fitting. "Technically successful but impractically expensive," Bruner told me. "We'll need live models for years yet."

Rita Mazzella is the doyenne of bra fit models. Lingerie companies use her breasts—size 34C—to construct new brassiere lines, then scale the size down to A and B, and up to D. "You probably think a model is a young, beautiful woman on a catwalk," she told me over lunch. "Not me. I'm a construction model. Designers put prototypes on me; I tell them when they get it right."

Mazzella, in her seventies, interrupted her packed schedule to meet me at a Madison Avenue café between modeling sessions. She told me she was born in Ponza, a small island off the Italian coast. When she was sixteen, her father, who had an import-export business, moved the family to New York, where she finished high school, attended the Fashion Institute of Technology, and was determined to become a model. An agency signed her, and sent her on her first job to model coats on Seventh Avenue. "Hated it," she said of having to dodge street-racks full of clothes outside and hustlers of all types inside. "A jungle."

Next the agency sent her to model lingerie on Madison Avenue. Those days, in the fashion industry, it was the far better side of the tracks. "For a female model, the intimate apparel business is fabulous. It's clean, people are respectful, nobody touches you, you're not asked to expose your body." She found the work interesting. "Outerwear doesn't have to do any functioning except drape right. A bra is complicated—a piece of engineering!—and all its parts have to work together." True Balance, the now-defunct firm she was sent to, liked the feedback she gave designers and hired her full time. She worked three years, quit to get married, and had a child. True Balance designers pleaded for her to return, and she worked freelance for that company and other agencies, quickly acquiring a reputation as a bra designer's dream.

The interaction of preference, performance, and fit makes bras next to impossible to standardize. "I once worked on a custom jean line with four customer preference styles," an industry specialist told me. "These ran from relaxed—that's the falling-off kind some teenagers wear—to ultra tight, with two levels in between. If the same person wore relaxed and ultra-tight, you know what the difference was? *Three inches!* Style and preference also make a difference with bras, though it's not so dramatic."

Mazzella owes her career to the fact that the two numbers used to characterize stock bra sizes—band length and cup size—are woefully inadequate to define a complex and mobile shape. This makes bra construction an engineering task requiring a standard, but also for female consumers makes the process of bra selection difficult. Internet sites that tell you how to self-size—which can ask for things like the number of fingers you can insert between your breasts or how many pencils your breast can hold underneath it without dropping—give laughably different results. Posture and shoulder shape affect bra style and size. So

Rita Mazzella receiving the 2002 Lifetime Achievement "Femmy" Award from the Underfashion Club.

does purpose: is it for comfort, jogging, enhancement, minimization, or a strapless dress?

Mazzella helps designers tailor their prototypes for those different purposes. "I know how to talk back," she said. "Fit models try thirty to forty bras a day, every day, week after week. After a while, your body becomes very sensitive, very attuned. You learn what's wrong when it doesn't fit, and what to tell the designer. That's not something a scanner can tell you. A scanner can't tell you what changes to make! A breast moves, it's alive!"

Fitting sessions can take just a few minutes, but Mazzella returns every day until she's satisfied. For most lines the process—fitting and refitting—takes a few days, but can take up to 3 months. One thing that often goes wrong is pitch. Pitch? Mazzella reaches for her glass and tilts it until the iced tea almost sloshes over the edge. "This glass has pitch. A bra can do the same thing; the balance can be off. What makes it do that? This is where a construction model helps."

Mazzella leaned over the table. "Sometimes I'll tell a designer, 'It's too tight here,' or 'The balance is now off' and she'll say, 'Rita, you're amazing! I just pulled it in an eighth of an inch!' That's what's valuable to companies. They could pull somebody off the street with perfect measurements, but they wouldn't be able to tell them that." In 2002, the Underfashion Club, a nonprofit support group for the intimate apparel business, gave Mazzella a Lifetime Achievement award, the first to a model, "in recognition of not only her long career in the intimate apparel industry, but for elevating the 'profile' of a fit model through her professionalism, grace and humor."

Still, Mazzella said, a lot of designing is now outsourced to China or done on mannequins. "When I first started my client book was *this* thick"—holding her thumb and forefinger 2 inches apart. "Dozens of mom and pop companies made bras. They've disappeared. Many that remain don't use live models. Only a few still do—Bali, Wacoal, Maidenform, Warners—I've worked for them all."

Mazzella had to get going. "Maidenform in New Jersey is having an emergency. They want a decision on a new line by the end of the day, and it takes an hour to get there. They need me to check it out before production." She laughed as she dashed off. "Sometimes I feel like a doctor!"

DARK SIDES OF THE METROSCAPE

Is the modern metroscape a utopia? Does it represent the culmination of how measurement can be optimally applied to giving us a grip on the world? Or does it represent the opposite; does measurement have a grip on us and obscure our grip on the world? The answer, alarmingly, may be "yes" to both. Are there, then, *other* ways of understanding ourselves and the world besides measurement and data collection?

One key difference between the modern metroscape and previous metroscapes is that ordinary life depends ever more on measurement, even as the management and intelligibility of the network is ever more removed from everyday life. While metrological matters were never really in the hands of the average citizen, comprehension generally was. Now, while use of the SI in everyday contexts might be easy, understanding its fundamentals is not, becoming too complex for all but scientists to grasp.

Weighing and measuring has always depended on circles of trust and expertise, but these circles have grown ever larger and more complex. The Bureau addresses the issue of trust by "traceability," or open publication of the calibrations and comparisons it has done of various standards, and by "mutual recognition agreements," which amount to certifications of trust by one national metrological institution in the performance of another. But these are reinforcements of trust within the metrological community and its institutions. Additional circles of trust have to connect these with the communities which use weights and measures.

I once saw a tape of a panel discussion at the Philoctetes Center, in New York City, entitled "Imposters, Forgery, Fraud, & Illusion," whose panelists included an author, an art conservation expert, a world-renowned magician, and a magic collector and criminal investigator. At one point, the criminal investigator illustrated the problem of trusting measurement systems by reporting on an experiment he had done in which he gave a group of people a series of unusual objects from Asia—an opium pipe, a bronze statuette, and a set of hara kiri knives—and asked them questions about the features of these objects, such as their age, origin, weight, and size. To describe their size, he allowed them to use a ruler that he conveniently produced. Then he asked the participants to rate their confidence level—1 to 100—in their answers. For age and origin the answers were in the 50 to 60 percent confidence level, but when the participants reported on the objects' size for which they had had the assistance of the ruler, they rated their confidence in the high 90s. However, the investigator had given them a ruler that was not 12 inches long but somewhat less than that. It had been made by unscrupulous lawyers—deliberately designed to deceive and appear to be 12 inches. When placed next to an object (a scar, a pothole, or the scene of an automobile accident) and photographed, a viewer would be inclined to believe the object to be larger than it really is, because that viewer is relying on a standard that he or she knows and trusts. The investigator's point is that we tend to trust the measurement network strongly: "Why would anyone ever question that this is not a ruler?" he asked as he revealed the deception to the audience. In the new metroscape, with the management and intelligibility of the system still more removed from the understanding of the average citizen, trust in the network becomes all the more important.

Other dangers lurk in the modern metroscape. From the beginning, it seemed to have a dark side associated with the potential for dehumanization and even sin. Kula mentions several pieces of European folk wisdom relating to the dangers of measuring: a memorable one is a piece of Czech folklore that children under six were not to be measured for clothes, lest they become runts or "measurlings," in the translator's brilliant rendering. Kula also cites the words of his Polish compatriot, the poet Adam Mickiewicz: "The compass, the scales, the yardstick—apply but to lifeless bodies."[8] Something about human life, Mickiewicz implies, escapes measuring, and while measuring sometimes tells us about ourselves and improves us, at other times it does not and even distracts us. Readers of Charles Dickens's novel *Hard Times* will remember Thomas Gradgrind, the dry, rational character who is "ready to weigh and measure any parcel of human nature, and tell you what it comes to," yet loses track of his own life.[9]

During the scientific revolution, one scientist after another—Galileo, Harvey, Kepler—made breakthroughs by simply measuring what he found. Area after area of human experience yielded its secrets to those who succeeded in finding a way to measure it. These stunning successes suggest that the real itself is the measurable. We measure because we assume that it can give us a better grip on the world. This assumption is deeply rooted in Western thought. In the *Republic*, Plato wrote that the best part of the human soul is the part which "puts its trust in measurement and calculating."[10] Yet, as German philosopher Martin Heidegger warned, the very successes of measurement can give us the impression that it is the *only* way to get a better grip on the world.

The chief feature of the modern era, which Heidegger called the

Gestell or "Enframing," is that the world entices us to measure it, and encourages us to think that there is no other way to find meaning in it.[11] We stop using measurement as a tool for understanding the world, and begin to see it as a tool for understanding ourselves. One scholar I know translates Heidegger's word informally as the "setup"; the world is set up nowadays so that measuring is the path to knowledge and understanding.[12] We have to be good at test-taking and measurement evaluation to navigate it successfully. The setup is not something subjective, in our heads; it is out in the world, something we encounter. But neither is it something objective, a natural part of the world, for it has arisen from the interaction of human beings with nature, it is something we have built into it, and it is wrapped up in our attitudes and the way we think and interact. It is therefore nothing from which we can escape. Usually taken for granted, the setup is noticeable wherever we get the sense that our measurement systems are running us, and that the reason we do things is because someone, somewhere, told us that this was the way.[13]

Modern literature and performance art is full of references to dehumanizing sides of measurement. In one terrifying scene in *Everything I Own I Carry With Me*, by the Romanian-born German novelist Herta Müller and based on a true story, the protagonist Oskar Pastior is imprisoned in a labor camp in the Soviet gulag, a brutal environment in which implacable measures—800 grams of bread daily, with his each shovelful of coal consuming 1 gram—govern life or death. Driven to delirium by hunger, he carries on hallucinatory conversations with what he calls the angel of hunger, who weighs each inmate on a gruesome scale:

> The angel of hunger looks at his scale and says: You are still not light enough for me—why don't you let go? I say: You turn my own flesh against me. I've lost it to you. But I am not my flesh. I am something

else, and won't let go. *Who* I am I can no longer say, yet neither can I say *what* I am. *What* I am is betrayed by your scale. . . .

One is not allowed to speak about hunger if one is hungry. Hunger is not like a bed frame, or it would have a measure. Hunger is not an object.[14]

In some regions of the modern metroscape, we demand exacting precision, while in others we are content with loose measurements and even prefer them. Consider sporting events where calls are relegated to the umpire, not to technology. Measuring a first down in football could easily be done far more precisely with the aid of a chip installed in the football and monitored by a GPS. But it would alter the accustomed momentum and spirit of the game.

Another peculiar region of the metroscape is in social science. According to Goodhart's law, whenever a measure is selected as a target for a particular policy it soon loses its value as a measure. Thus if a school board designates higher SAT scores as a target for improving schools, or if a nation designates a higher GNP as a measure of social welfare, ways to boost that measure are found that do not affect the goal they mean to attain, and thus the measure ceases to be a measure. Still another region is at the atomic level, where in the famous problem of quantum measurement the way we set up measuring equipment affects the outcome; measurement changes what is measured in uncanny ways.

In the past, the social context of measuring was fairly apparent to users of the system: the Akan measurers of gold were aware of the nuances of their activity, the members of the Chinese court were attuned to the politics of precision, and the peasants of premodern Europe were all too painfully aware of the potential for exploitation in the way they measured. In the modern metroscape, this is less obvious. The social impact of using measuring to establish intelligence, say, or the impact of educational institutions, is more disguised and involves demoting the importance of anything that does not involve measurement, and augmenting the importance of anything that does.

The modern metroscape, illustrated by the contrast between Vitruvian Man and Joe and Josephine, involves a new relation between measurement and how we relate to the world and each other. This metroscape—which tends to hide but can be brought to light—shapes what we make, what we purchase, how we classify things, and what we consider real. It is far from being a utopia, but by understanding it we can prepare for the dangers it conceals.

CHAPTER TWELVE

AU REVOIR, KILOGRAM

The event at the Royal Society in London, on January 24, 2011, began precisely on time. After the last delegates had taken their seats, Stephen Cox, the institution's executive director, noted sheepishly that the wall clock was running "a little slow" and promised to reset it. Cox knew the audience cared about precision and would appreciate his vigilance. As the world's leading metrologists, they had gathered to discuss a sweeping reform of the scientific basis of the International System of Units—the SI—in the most comprehensive revision yet of the international measurement structure that underpins global science, technology, and commerce.

If approved, these changes would involve redefining the seven SI base units in terms of fundamental physical constants or atomic properties. The most significant of these changes would be to the kilogram, a unit that continues to be defined by the mass of a platinum-iridium cylinder at the BIPM outside Paris, and the only SI unit still defined by an artifact. Metrologists want to make these changes for several reasons, including worries about the stability of the kilogram artifact, the need for greater precision in the mass standard, the availability of new technologies that seem able to provide greater long-term precision, and the desire for stability and elegance in the SI's structure.

Over the 2 days of the meeting, participants expressed varied opinions about the force and urgency of these reasons. One of the chief enthusiasts and prime movers behind the proposed changes was former BIPM director Terry Quinn, who also organized the meeting. "This is indeed an ambitious project," he said in his opening remarks. "If it is achieved, it will be the biggest change in metrology since the French Revolution."[1]

THE METRIC SYSTEM AND THE SI

The French Revolution, as we saw in Chapter 4, had indeed brought about the single greatest change in metrology in history. Instead of merely reforming France's unwieldy, inherited weights and measures, vulnerable to error and abuse, the Revolutionaries imposed a rational and organized system, devised by the *Académie des Sciences* and intended "for all times, for all peoples," that tied length and mass standards to natural standards: the meter to one forty-millionth of the Paris meridian, the kilogram to the mass of a cubic decimeter of water. But maintaining the link to natural standards proved impractical, and almost immediately length and mass units of the metric system were enshrined instead, using the artifacts deposited in the National Archives in 1799. The big change now being championed was to achieve at last what was aimed at in the eighteenth century, to base standards on constants of nature.

Despite its simplicity and rationality, the metric system took decades to implement in France. It also spread slowly beyond French borders, though other nations eventually began to adopt it out of a mixture of motives: fostering national unity, repudiating colonialism, enhancing competitiveness, and as a precondition for entering the world community. In 1875, as we saw, the Treaty of the Meter removed supervision of the metric system from French hands and assigned it to an international body, the BIPM. The treaty also initiated construction of new length and mass standards—the International Prototype of the Meter and the International Prototype of the Kilogram—to replace the meter and kilogram made by the Revolutionaries. These were manufactured in 1879 and officially adopted in 1889–but they were calibrated against the old Meter and Kilogram of the Archives.

At first the BIPM's duties primarily involved caring for the prototypes and calibrating standards of member states. But in the first half of the twentieth century, it broadened its scope to cover other kinds of measurement issues as well—including electricity, light, and radiation—and expanded the metric system to incorporate the second and ampere in the so-called MKSA system. Meanwhile, advancing interferometer technology, from Michelson and Morley onward, allowed scientists to measure length with a precision rivalling that of the meter prototype. In 1960, as we saw in Chapter 10, these developments culminated in two far-reaching changes, made at the Eleventh General Conference on Weights and Measures (CGPM), the meeting of member states that ultimately governs the BIPM and meets every 4 years. The first was to redefine the meter in terms of the light from an optical transition of krypton-86. (In 1983, the meter would be redefined again, in terms of the speed of light.) No longer would nations have to go to the BIPM to calibrate their length standards; any country could realize the meter, provided it had the technology. The International Prototype of the Meter was relegated to a historical curiosity; it remains in a vault at the BIPM today.

The second revision of the 1960 meeting of the CGPM was to replace the expanded metric system with a still greater framework for the entire field of metrology. The framework consisted of six basic units—the meter, kilogram, second, ampere, degree Kelvin (later the kelvin), and candela (a seventh, the mole, was added in 1971), and a set of "derived units," such as the newton, hertz, joule, and watt, built from these six. Because it amounted to such a significant change from what came before, it was given a new name and baptized the International System of Units. But the SI, as it was known after its French initials, still based the kilogram definition on an artifact, the International Prototype of the Kilogram manufactured in 1879.

This 1960 reform was the first step toward the current overhaul. More steps soon followed. With the advent of the atomic clock and the ability to measure atomic processes with precision, the second was redefined in 1967 in terms of nuclear properties, called hyperfine levels, of caesium-133. The strategy once again involved scientists measuring a fundamental property with precision, and then "bootstrapping," that is, redefining the unit in which the property was measured in terms of a fixed value of that property. The property then ceased to be measurable within the SI, and instead defined the unit.

The kilogram, however, stubbornly resisted all attempts to redefine it in terms of a natural phenomenon: mass proved exceedingly difficult to scale up from the micro- to the macroworld. But because mass is involved in the definitions of the ampere and mole, this brought to a halt attempts for more such redefinitions of units.

Metrologists discussed this predicament in 1975, when several hundred scientists gathered in Paris to celebrate the 100th anniversary of the Treaty of the Meter. The event was a conference called "Atomic Masses and Fundamental Constants." Participants were treated to a reception at the Elysée Palace hosted by French Prime Minister Valéry Giscard d'Estaing, and to a lavish conference banquet at the fabulously scenic Jules Verne restaurant at the Eiffel Tower. BIPM director Jean Terrien told conference participants that they had much to celebrate: "For the first time in the history of humanity, a single system of units is accepted by the entire world."[2]

By 1975, indeed, virtually all the major countries on Earth had switched to a mandatory use of the SI. The one significant holdout was the United States, but it seemed like the country was about to embark on a serious effort to convert. ("Drugs have taught an entire generation of Americans the metric system," the author P. J. O'Rourke quipped, but that was as far as many Americans whose careers lay outside sciences got.) Four years earlier, the U.S. National Bureau of Standards had given Congress a report called "A Metric America: A Decision Whose Time Has Come," outlining the reasons for converting to the metric system. Congress had passed one pro-metric piece of legislation in 1974, promoting its use in education, and in 1975 was debating a "Metric Conversion Act." Passage seemed likely, and indeed the bill was passed and signed into law by the end of the year. The United States, it appeared, was finally about to switch to the metric system.

Terrien also mentioned that some units had been tied to natural standards—but he had to admit that the kilogram was the last remaining artifact standard and that replacing it with a natural standard was still a "utopian" dream. It looked like the International Prototype of the Kilogram (IPK) was here to stay.

DRIFTING STANDARD

A pivotal event took place in 1988, when the IPK was removed from its safe and compared with the six identical copies kept with it, known as *témoins* (witnesses). The previous such "verification," which took place in 1946, had discovered slight differences among these copies, attributable to chemical interactions between the surface of the prototypes

THE INTERNATIONAL PROTOTYPE OF THE KILOGRAM, HOUSED IN A SAFE IN THE BASEMENT OF THE INTERNATIONAL BUREAU OF WEIGHTS AND MEASURES (BIPM) OUTSIDE PARIS.

and the air, or to the release of trapped gas. But the verification in 1988 yielded a surprise: the masses of the témoins appeared to be drifting upward with respect to that of the prototype. The verification in 1988 confirmed this trend: not only the masses of the témoins but those of practically all the national copies had drifted upward with respect to that of the prototype, the prototype differing in mass from them by about +50 μg, or a rate of change of about 0.5 parts per billion per year. The international prototype behaved differently, for some reason, from its supposedly identical siblings.

Quinn, who became the BIPM's director in 1988, outlined the worrying implications of the apparent instability of the IPK in an article published in 1991.[3] Because the prototype *is* the definition of the kilogram, technically the témoins are gaining mass. But the "perhaps more probable" interpretation, Quinn wrote, "is that the mass of the international prototype is falling with respect to that of its copies"; that is, the prototype itself is unstable and losing mass. Although the current definition had "served the scientific, technical, and commercial communities pretty well" for almost a century, efforts to find an alternative, he suggested, should be redoubled. Any artifact standard will have a certain level of uncertainty because its atomic structure is always changing—in some ways that can be known and predicted and therefore compensated for, in others that cannot. Furthermore, the properties of an artifact vary slightly with temperature. The ultimate solution would be to tie the mass standard, like the length standard, to a natural phenomenon. But was the technology ready? The sensible level of accuracy needed to replace the International Prototype of the Kilogram, Quinn said, was about 1 part in 10^8.

In 1991, two remarkable technologies—each developed in the previous quarter century, and neither invented with mass redefinition in mind—showed some promise of being able to redefine the kilogram. One approach, the "Avogadro method," realizes the mass unit using a certain number of atoms via the construction of a sphere of single-crystal silicon and a measurement of the Avogadro constant. The "watt balance" approach, on the other hand, ties the mass unit to the Planck constant, via a special device that exploits the equality of SI units of mechanical and electrical power. The two approaches are comparable because the Avogadro and the Planck constants are linked via other constants whose values are already well measured, including the Rydberg and fine-structure constants. Although neither approach was, in 1991, close to being able to achieve a precision of a part in 10^8. Quinn thought at the time that it would not be long before one or both would be able to do this. His optimism was misplaced.

THE SPHERE . . .

The Avogadro approach links micro- and macroscales by defining the mass unit as corresponding to a certain number of atoms using the Avogadro constant, which relates the number of elementary entities (atoms, say) to the molar mass of a substance—traditionally the number of atoms of carbon, whose atomic mass is 12, in 12 grams of carbon—and is about 6.022×10^{23} mol^{-1}. It would, of course, be impossible to count that many atoms one by one, but it can be done instead by making a perfect enough crystal of a single chemical element and knowing the isotopic abundances of the sample, the crystal's lattice spacing, and its density. Silicon crystals are ideal for this purpose as they are produced by the semiconductor industry to high quality. Natural silicon has three isotopes—silicon-28, silicon-29, and silicon-30—and initially it seemed their relative proportions could be measured sufficiently accurately. Although measuring lattice spacing proved harder, metrologists drew on a technique called Combined Optical and X-ray Interferometers (COXI), which was pioneered in the 1960s and 1970s at the German national-standards lab (PTB) and the U.S. National Bureau of Standards (NBS)—the forerunner of the National Institute of Standards and Technology (NIST). It relates X-ray fringes—hence metric length units—directly to lattice spacings. The trick was to use the X-rays to measure a Moiré interference pattern—one formed by the overlay of two different wave patterns—created by three thin crystal layers called lamella. When one of these lamella is synchronized with an optical laser interferometer and then slowly moved, it is possible to link the spacings in the Moiré pattern with an optical laser frequency. For a time in the early 1980s, the results of the two groups differed by a full part per million. This disturbing discrepancy was finally explained by an alignment error in the NIST instrument, leading to an improvement in the understanding of how to beat back systematic errors in the devices.

The source of uncertainty that turned out to be much more difficult to overcome involved determination of the isotopic composition of silicon. This appeared to halt progress toward greater precision in the measurement of the Avogadro constant at about 3 parts in 10^7. Not only that but the first result, which appeared in 2003, showed a difference from the watt balance results of more than 1 part in a million (1 ppm). There was a strong suspicion that the difference stemmed from the measurements of the isotopic composition of the natural silicon used in the experiment. The leader of the PTB team, Peter Becker, then had a stroke of luck. A scientist from the former East Germany, who had connections to the centrifuges that the Soviets had used for uranium separation, asked Becker if it might be possible to use

A HIGH-PRECISION SILICON SPHERE, TO BE USED IN ONE APPROACH TO DEFINING THE KILOGRAM IN TERMS OF PLANCK'S CONSTANT.

enriched silicon. Realizing that using a pure silicon-28 sample would eliminate what was then thought to be the leading source of error, Becker and collaborators jumped at the opportunity. Although buying such a sample would be too costly for a single lab—an eye-watering €2 million for 5 kilograms of the material—in 2003, representatives of Avogadro projects from around the world decided to pool resources to buy the sample and form the International Avogadro Coordination (IAC). Becker at the PTB managed the group, parcelling out tasks such as characterization of purity, lattice spacing, and surface measurements to other labs.

Two beautiful spheres resulted. "It looks like what we've made is just another artifact like the kilogram—what we are trying to get away from," Becker said at the January Royal Society meeting. "It's not true—the sphere is only a method to count atoms."[4]

. . . AND THE BALANCE

The second approach to redefining the kilogram involves an odd sort of balance. Whereas an ordinary balance compares one weight against another—a bag of apples, say, versus something else of known weight—the watt balance matches two kinds of forces: the mechanical weight of an object with the electrical force of a current-carrying coil placed in a strong magnetic field.[5]

What is remarkable about the watt balance is how it relies on several astonishing discoveries, none of which were made by scientists attempting mass measurements. One is the Josephson effect, which can measure voltage precisely. The next is the quantum Hall effect (discovered by Klaus von Klitzing in 1980), which can measure resistance precisely. The third is the concept of balancing mechanical and electrical power, which can be traced back to Bryan Kibble at the NPL in 1975, who had actually been

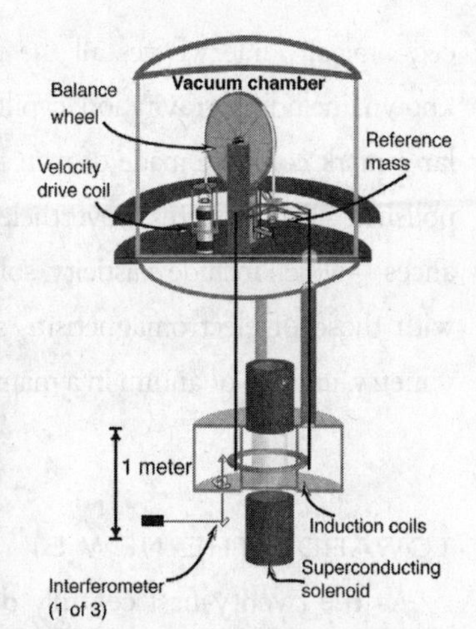

The watt balance, a device that links the kilogram with Planck's constant, *h*, allowing the kilogram to be defined in terms of a natural constant. A kilogram test mass is placed on a balance pan connected to a coil of copper wire surrounding a superconducting electromagnet. If electric current is sent through the coil, then just as in an electric motor, electromagnetic forces are produced to balance the weight of the test mass. The apparatus measures this current and force. The apparatus can move the coil vertically also, which induces a voltage as in an electric generator, and the velocity and voltage of the coil are measured. These four measurements determine the relationship between mechanical and electrical power, which can be combined with other basic properties of nature to redefine the kilogram in terms of Planck's constant.

trying to measure the electromagnetic properties of the proton.[6] These three discoveries can now be linked in such a way that the kilogram can be measured in terms of the Planck constant, a number that reflects the size of the smallest amount of energy that can exist independently. By "bootstrapping," the process could now be in principle reversed, and a specific value of the Planck constant used to define the kilogram.

In Michael Faraday's famous talk, "A Chemical History of the Candle," the British scientist called candles beautiful because their operation economically interweaves all the fundamental principles of physics then known, including gravitation, capillary action, and phase transition. A similar remark could be made of watt balances, which though not as pretty as polished silicon spheres, nevertheless combine the complex physics of balances—which include elasticity, solid state physics, and even seismology—with those of electromagnetism, superconductivity, interferometry, gravimetry, and the quantum in a manner that exhibits deep beauty.

TOWARD "THE NEW SI"

As the twenty-first century dawned, both the Avogadro approach and watt balances had reached an accuracy of about a few parts in 10^7, still far from Quinn's target of 1 part in 10^8. Nevertheless Quinn, who stepped down as BIPM director in 2003, decided to pursue the redefinition. Early in 2005, he coauthored a paper entitled "Redefinition of the Kilogram: A Decision Whose Time Has Come," the subtitle derived from the (by then ironic) NBS report of the 1970s heralding imminent U.S. conversion to the metric system. (The United States never converted and appears unlikely to anytime soon, given the new ease, thanks to computers, of converting imperial to metric units, and U.S. politicians' fear of embarking on reform.) "The advantages of redefining the kilogram immediately outweigh any apparent disadvantages," the authors wrote, despite the then apparent inconsistency of 1 ppm between the watt balance and silicon results.[7] They were so confident of approval by the next CGPM—the twenty-third in 2007—that they inserted language for a new definition into "Appendix A" of the BIPM's official SI brochure. Furthermore, they wanted to define each of the seven SI base units in terms of physical constants or atomic properties. In February 2005, Quinn organized a meeting at the Royal Society to acquaint the scientific community with the plan.

SI AND THE NEW SI: DEFINITION OF THE SEVEN BASE UNITS

Base Quantity	Base unit	Symbol	Reference constants used to define the unit in current SI	Reference constants used to define the unit in "New SI"
time	second	s	hyperfine splitting in Cs-133	hyperfine splitting in Cs-133
length	meter	m	speed of light in a vacuum *c*	speed of light in a vacuum *c*
mass	kilogram	kg	mass of International Prototype Kilogram	Planck constant *h*
electric current	ampere	A	permeability of free space	elementary charge *e*
thermodynamic temperature	kelvin	K	triple point of water	Boltzmann constant *k*
amount of substance	mole	mol	molar mass of carbon-12	Avogadro constant N_A
luminous intensity	candela	cd	luminous efficacy of a 540 THz source	luminous efficacy of a 540 THz source

The reaction ran from lukewarm to hostile. "We were caught off guard," as one participant put it. The case for the proposed changes had not been elaborated at the 2005 meeting, and many thought it unnecessary, given that the precision available with the existing artifact system was greater than that of the two newfangled technologies. Not only were the uncertainties achieved by the Avogadro and watt balance approaches at least an order of magnitude away from the target Quinn had set in 1991, but there was still this difference of 1 ppm to be accounted for. Nevertheless, the idea had taken hold, and the BIPM's governing board, the International Committee for Weights and Measures, in October 2005 adopted a Recommendation in which it envisaged a redefinition of the kilogram as it had been proposed in the 2005 paper but it went further and included redefinitions of four base units (kilogram, ampere, kelvin, mole) in terms of fundamental physical constants (the Planck constant *h*, the elementary charge *e*, Boltzmann's constant *k*, and the Avogadro constant N_A, respectively). Quinn and his colleagues then published a second paper in 2006 in which they gave specific proposals designed to implement the CIPM Recommendation, but not for 2007. Instead they proposed that the twenty-fourth General Conference in 2011 consider a resolution expressing the intent to implement the recommendation pending further improvements in data and technology.

Since 2006 both approaches have realized significant progress. In 2004, enriched silicon in the form of SiF_4 was produced in St. Petersburg and converted into a polycrystal at a lab in Nizhny Novgorod. The polycrystal was shipped to Berlin, where a 5-kg rod made from a single crystal of silicon-28 was manufactured in 2007. The rod was sent to Australia to be fashioned into two polished spheres, and the spheres were measured in Germany, Italy, Japan, and the BIPM. In January 2011, the IAC reported a new measurement of their results with an uncertainty of 3.0×10^{-8}, tantalizingly close to the target. The result, the authors write, is "a step towards demonstrating a successful *mise en pratique* of a kilogram definition based on a fixed N_A or h [Avogadro or Planck] value" and claim it is "the most accurate input datum for a new definition of the kilogram."[8]

Watt balance technology, too, has been steadily developing. Devices with different designs are under development in Switzerland, France, China, Canada, and the BIPM. The results indicate the ability to reach an uncertainty of less than 10^{7}. Their principal problem is alignment: the force produced by the coil and its velocity must be carefully aligned with gravity. The more the overall uncertainty is reduced, the more difficult it is to make these alignments. The previous difference of 1 ppm has been reduced to about 1.7 parts in 10^{7}, close but not quite close enough.

These results changed attitudes toward the proposed alterations, leading to a near-consensus in the metrological community that a redefinition is not only possible but likely. A BIPM advisory committee has proposed criteria for redefinition: there should be at least three different experiments, at least one from each approach, with an uncertainty less than 5 parts in 10^{8}; at least one should have an uncertainty less than 2 parts in 10^{8}; and all results should agree within a 95 percent confidence level. At present the new technologies will not provide greater overall precision, but will provide greater stability. In the long run, the uncertainty of watt balances is very likely to drop below a part in 10^8 to a few parts in 10^9. At present masses can be compared with a precision of a part in 10^{10}, but the best uncertainties are a few parts in 10^9 in relating mass to the International Prototype of the Kilogram.

The proposal drafted by Quinn and colleagues, for what is called the "New SI," did not actually redefine the kilogram but took note of the intention to do so. Their proposal bundled the redefinition of the kilogram together with the redefinitions of the other base units in terms of constants in a single package. The kilogram redefinition is the most problematic, requiring more technological development and scientific data, and its redefinition is also tied to that of the ampere and mole. The redefinition of the kelvin in principle could be pursued independently, but the BIPM's officials thought it best, for educational and managerial reasons, to make one coordinated change in the world's metrological system rather than changing it in pieces.

SI VS. NEW SI DEFINITIONS OF THE KILOGRAM

SI DEFINITION: "The kilogram is the unit of mass; it is equal to the mass of the international prototype of the kilogram. It follows that the mass of the international prototype of the kilogram, $m(\mathbf{K})$, is always 1kg exactly."

NEW SI DEFINITION: "The kilogram, kg, is the unit of mass; its magnitude is set by fixing the numerical value of the Planck constant to be equal to exactly 6.626 068 . . . $\times 10^{-34}$ when it is expressed in the unit s^{-1} m^2 kg, which is equal to J s." The ellipses (. . .) indicate that the exact number is yet to be specified. The definition also leaves open the technology (*mise en pratique*) by which this definition is to be realized.

These developments gave Quinn the confidence to organize another meeting. This time, he and his fellow organizers devised a careful strategy. The 150 participants at the January 2011 Royal Society meeting included three Nobel laureates: John Hall of JILA (whose work contributed to the redefinition of the meter), Bill Phillips of NIST, and von Klitzing himself. No longer will these physical constants be measured because, within the SI, their numerical values are fixed. Furthermore, the definitions are similar in structure and wording, and the connection to physical constants made explicit. The language makes clear what these definitions really mean—what it *means* to tie a unit to a natural constant. The definitions therefore have a conceptual elegance.

A SINGLE STATEMENT TO DEFINE THE ENTIRE NEW SI:

The International System of Units, the SI , is the system of units in which

- the ground state hyperfine splitting frequency of the caesium 133 atom $\Delta\nu(^{133}\mathrm{Cs})_{\mathrm{hfs}}$ is exactly 9 192 631 770 hertz,
- the speed of light in vacuum c is exactly 299 792 458 metre per second,
- the Planck constant h is exactly 6.626 068 ... $\times 10^{-34}$ joule second,
- the elementary charge e is exactly 1.602 176 ... $\times 10^{-19}$ coulomb,
- the Boltzmann constant k is exactly 1.38065 ... $\times 10^{-23}$ joule per kelvin,
- the Avogadro constant N_A is exactly 6.022141 ... $\times 10^{23}$ reciprocal mole,
- The luminous efficacy K_{cd} of monochromatic radiation of frequency 540×10^{12} Hz is exactly 683 lumen per watt.

The metrological community is vast and diverse, and different groups tend to have different opinions about the proposals. Those who make electrical measurements tend to be enthusiastic; the Planck constant and the value of the elementary charge, which such scientists use constantly, now became exactly determined and much easier to work with. Moreover, the awkward split between the best available electrical units introduced by Kibble's device and those available in the

SI is eradicated. The only outright objection from this corner at the meeting, by von Klitzing, was tongue-in-cheek. "Save the von Klitzing constant!" he protested, pointing out that his eponymous constant, which has been conventionally set (outside the SI) at 25812.807 ohms exactly for two decades, now becomes revalued, making it long and unwieldy rather than short and neat. Yet he went on to express sympathy with the redefinition, citing Max Planck's remark that "with the help of fundamental constants we have the possibility of establishing units of length, time, mass, and temperature, which necessarily retain their significance for all cultures, even unearthly and nonhuman ones."[9]

The mass-measurement community tends to be less sanguine. Mass measurers can currently compare masses with about an order of magnitude greater precision—a part in 10^9—than they can achieve by directly measuring a constant. The new definitions thus appear to introduce more uncertainty into mass measurements than exists at present; in place of careful traceability back to a precisely measurable mass, one now has traceability back to a complicated experiment at various national laboratories. As Richard Davis, former head of the BIPM mass division, remarked about the SI, "It's got to be like a piece of Shaker furniture: not just beautiful but functional." Advocates, however, point out that comparison measurements conceal the uncertainty present in the kilogram artifact itself, so that ultimately no new uncertainty is introduced. "Uncertainty is conserved," as Quinn remarked.

One interested group not present at the meeting were students, educators, and other members of the public interested in understanding metrology. As the *Chicago Daily Tribune* had complained after the SI was created in 1960, fundamental measurement issues, which should be simple for the average person to understand, are becoming too complex for anyone except scientists. One of the attractions of science for students is that the concepts and practices are perspicuous, or aim to be—but the new SI seems to put the foundations of metrology out of reach for all but insiders. Woe to butchers and grocers, someone may jest when the new SI takes effect, who are not proficient in quantum mechanics.

Still, each age bases its standards on the most solid ground it knows, and it is appropriate that in the twenty-first century this includes the quantum, a fundamental piece of the structure of the world as we know it. None of the attendees at the Royal Society meeting objected in principle to the idea that the kilogram eventually should be redefined in terms of the Planck constant. "It's a scandal that we have this kilogram changing its mass—and therefore changing the mass of everything else in the universe," Phillips remarked at one point.[10] A few people were bothered that it now appeared impossible for scientists to detect whether certain fundamental constants were changing their values, though others pointed out that such changes would be detectable by other means. Many participants, however, were troubled that the Avogadro and watt balance teams have produced two measurements that are still not quite in sufficiently good agreement, throwing a monkey wrench, if not perhaps a large one, into the attempt to pick a single value. "The person who has only one watch knows what time it is," said Davis, citing an ancient piece of metrology wit; "The person who has two is not sure."

Quinn was confident that the discrepancy will be resolved in a few years. The only clear controversy on view at the meeting concerned what would happen if it is not. Quinn wanted to plunge ahead with the redefinition anyway, given that the discrepancy involves levels of precision so tiny that they would not affect measurement practice. Some objected, fearing that there would be secondary effects, such as in legal metrology, which concerns legal regulations incorporated into national and international agreements (when someone suggested that metrologists would easily be able to persuade lawyers to revise their documents in the light of the cold data of science, titters swept the room). Others worry about the perception of metrology and metrological laboratories if a value is fixed prematurely and then the mass scale must be changed in the light of better measurements in the future. NPL director Brian Bowsher, referring to the current climate change controversy, in which skeptics leap on any hint of uncertainty in measurements, stressed the importance of being "the people who take the time to get it right."

Calling the New SI the greatest change since the French Revolution may be hyperbole. The advent of the SI in 1960 was certainly radical in that it introduced new units and tied existing ones to natural phenomena for the first time. The new changes, too, will have scarcely any impact on measurement practice, and are largely for pedagogical and conceptual reasons. The New SI does, however, represent a transformation in the status of metrology. When the SI was created by the Eleventh CGPM in 1960, metrology was regarded as something of a backwater in science, almost a service occupation. Metrologists built the stage on which scientists acted. They provided the scaffolding—a well-maintained set of measuring standards and instruments, and a well-supervised set of institutions that cultivated trust—that enabled scientists to conduct research. The New SI, and the technologies that make it possible, connect metrology much more intimately with fundamental physics.

On October 21, 2011, the twenty-fourth CGPM unanimously adopted a resolution expressing the intent to implement the New SI. A few lines were added to the resolution asking the CIPM to carry out further efforts to make everything understandable as far as possible to the general public, but maintaining scientific rigor and clarity. The resolution did not actually redefine the kilogram in terms of h, but stipulated that a final decision would only be made when sufficient consistent and accurate data existed for the scientists to agree on an accepted value for h.

The New SI resembles the noonday cannon in the loose sense that its standards are still conventional, a result of human choice. There remains a final turtle, so to speak, in our measuring system; actually, seven of them, one for each of the seven base units. But these final turtles—physical constants or atomic properties—are the product of the scientifically and technologically rich twenty-first century environment, and more abstract than ever before, more the product of human planning and deliberation. These standards are also woven more intimately into the fabric of the world's structure than any previous ones. The New SI will certainly come to be supplemented with more derived SI units to suit changes in science, and be accompanied by more off-SI units that spring up spontaneously to suit the purposes of everyday life. It may be further revised in the future. But the new metrological system is connected as never before with our best judgment of the ultimate structure of the physical world. It is breathtaking in its ambition, and the realization of a centuries-old dream.

EPILOGUE

Inchworm, inchworm
Measuring the marigolds
You and your arithmetic
You'll probably go far

Inchworm, inchworm
Measuring the marigolds
Seems to me you'd stop and see
How beautiful they are

—Frank Loesser

There are two very different ways of measuring, the Greek philosopher Plato pointed out. One, the kind discussed in this book, involves numbers, units, a scale, and a beginning point. It establishes that one property is greater than or less than another, or it assigns a number to how much of a given property something possesses. We might call this "ontic" measuring, after the word applied by philosophers to real independently existing objects or properties. This book has related the story of how ontic measurement developed from improvised body measures and disconnected artifacts into a single, universal network that relates many different kinds of measurements and ties them all ultimately to absolute standards—physical constants.

Yet another kind of measuring does not involve placing oneself next to a stick or in a pan. This is the kind of measurement that Plato said is guided by a standard of the "fitting" or the "right." This kind of measuring is less an act than an experience; the experience that things that we've done, or we ourselves, are less than they could or should be. We cannot carry out this kind of measuring by following rules, and it does not lend itself to quantification. Is this only "metaphorical" measuring? It is comparison against a standard. Placed alongside the fitting or the right example, our actions—and even our *selves*—do not have enough being; there is more to be. We feel we are not measuring up to our potential. We might call this "ontological" measuring, after the word philosophers use to describe the way that something exists.

Ontological measuring involves no specific property, in a literal-minded respect, for it involves nothing quantitative. Calculate all we please, we will never produce this kind of measurement. No method can lead us to it. Ontological measurement connects us with something trans-human, something *in* which we participate, not something *over* which we command. While in ontic measurement we compare some object with another object exterior to it, in ontological measurement we compare ourselves, or something we have produced, with something in which our being is implicated, to which it is related—such as some concept of the good, the just, or the beautiful. Ontological measurement is ontically measureless.

For religious believers, we humans ordinarily lead imperfect lives: we don't measure up. Scholars of the ancient world were mostly confident that standards for our potential existed and that human beings could find such standards and use them as measures. Aristotle described the moral man as a "measure" in Book 9 of his *Nicomachean Ethics*. By this he meant not that a moral man is something against which we can physically or even symbolically compare ourselves, but that our encounters with genuinely moral human beings "call us out of ourselves," often quite forcefully, making us want to be better humans.

Ontological measuring is the measuring that good examples invite. The history of literature and art is replete with great works and performances that each artist can experience as intangible yardsticks for measuring his or her own achievements. These great works and performances of the past scream out to artists of the present that there is still more to do. Traditions change, and with them ideas of what is good and what not. But traditions provide an authenticating horizon in which artists experience a measure of what is good and what not, what original and what an echo, what vibrant and full of life and what deficient.

The way philosophers often describe the "call of conscience" involves ontological measurement, a secular variation on the old spiritual idea of being "called back to yourself." Conscience, like other forms of ontological measurement, requires opening ourselves to being able to say, "I could be better," opening ourselves to being able to experience ourselves as ontologically deficient—a positive thing![1] This is the foundation of ethics. When human beings lack that sense—and understand ethics to consist in, say, simply obeying rules—such behavior has little to do with genuine ethics. For why do we choose *these* rules, and know that they are in fact *moral* rules? Because someone told us so? It is only because we *already* know what ontological deficiency is like—we've experienced it—and *already* know that these rules help guide us toward the kind of behavior that tends to better those who follow them.

Artists and performers especially—but also engineers, educators, businesspeople, judges, and those of other professions—practice both kinds of measuring all the time. But the two kinds of measurements are often confused, often with damaging results. Stephen Jay Gould's book *The Mismeasure of Man* is about the fallacy that "worth can be assigned to individuals and groups by measuring intelligence as a single quantity." Shakespeare's play *Measure for Measure*—an allusion to Matthew 7:1–2: "Judge not, that you be not judged. For with the judgment you pronounce you will be judged and the measure you give will be the measure you get"—is about the need to temper the literal application of legal measures with empathy and mercy in order to live up to what it means to be fully human. Moral thinking begins with the distinction between ontic and ontological measures.

Heidegger was fond of citing the poet Friedrich Holderlin's often-quoted passage: "Is there on earth a measure? There is none."[2] If so, it reflects an odd state of affairs: as the modern world has progressively improved and perfected its ontic measures, it has diminished its ability to measure itself ontologically. How could this have happened?

The reason is that ontic measuring can distract from, and even have a corrosive effect on, ontological measurement. The ancient scholar Flavius Josephus, referring to Adam's child Cain, said that "He also introduced a change in that way of simplicity wherein men lived before; and was the author of measures and weights. And whereas they lived innocently and generously while they knew nothing of such arts, he changed the world into cunning craftiness."[3] Something about the practice of measurement, Josephus implied, tends to make cheaters of us all. We were better before its introduction.

The capacity and new tools for measurements in our lives seems continually on the increase and can appear to be an unqualified good. A Web site, The Quantified Self, bills itself as providing "tools for knowing your own mind and body." These tools are means for collecting data about the times we spend in such activities as working, eating, sleeping, having sex, worrying, cleaning up, having coffee, and every other aspect of everyday life. "Behind the allure of the quantified self," wrote the site's cofounder in *The New York Times*, "is a guess that many of our problems come from simply lacking the instruments to understand who we are."[4] How fortunate, therefore, that we are to be able to quantify every aspect of our lives in this high-speed, rapidly changing world! No ambiguity here. Measurement is an indispensable tool of self-knowledge. The better we do it, the more we know of our selves.

By contrast, *Vital Statistics of a Citizen, Simply Obtained*, a 40-minute video of a performance piece by the American artist Martha Rosler (1977), depicts measurement as utterly dehumanizing. Most of the video consists of a thirty-three-year-old woman being measured by two white-coated men, one of whom makes the measurements while the other writes them down. At first they stand her against the wall and draw a Vitruvian Man–like measured image of her with outstretched limbs. Then they ask her to take off more of her clothes as they measure more intimate parts of her body, culminating in her "vaginal depth." They have her lie down horizontally in front of her measured image. As she is being measured and the one male announces each of her measures to the other, he calls it "below standard" (whereupon the soundtrack has a razzing sound), "above standard" (a beep), or "standard" (pleasant chimes). Meanwhile, a feminine voiceover characterizes what's happening in apocalyptic terms, referring to rape, dehumanization, degradation, exploitation, eugenics, and tyranny; the voiceover says that the woman is being indoctrinated to manage her image, to view her body as parts, and to lose track of her self, and quotes the philosopher Jean-Paul Sartre to the effect that "Evil is the product of the ability of humans to make abstract that which is concrete." After the male measurers have their way with the woman, she puts clothes back on in two sequences: in one she dons a wedding dress and veil, in the other a hot little black dress. The wedding dress sequence has her returning to the wall to stand demurely and compliantly next to her measured image; in the black dress sequence she darts off in the opposite direction. The video ends by returning to the two male doctors summoning another woman: "Next!"[5]

No ambiguity here, either. Measuring does something bad to us. What is wrong is not simply misplaced precision; too much of a good thing. Measuring is far more sinister, a tool of oppression. It destroys our selves, or at least those of women; men evidently either had no selves to begin with or shed theirs long ago. Best for those who still have selves to renounce measurement.

The "setup" means that the environment in which we measure is not neutral; this was Heidegger's point. In the modern atmosphere, measuring tends to dazzle and distract us. We tend to look away too much from what we are measuring, and why we are measuring, to the measuring itself. Measuring certainly works and helps us to get around, but in the modern metroscape it can lead us to think that it is all we need to get around. Policy questions or key decisions—"Should we fire our teachers?" "What college should I attend"—are decided by measurement questions—"What are the SAT scores?" "What are the rankings?"

The Vitruvian Man was an ideal image, something that connected human beings with beauty, perfection, and other transhuman goals, goals toward which measures could at best only serve as signposts. Joe and Josephine are something different; they are models, means for designers to achieve the efficient creation of interfaces between human beings and the world. Transhuman goals are absent; Joe and Josephine assist the aim of putting the world at the disposal of the wants and needs of human beings. My [TC]² avatar is still more remote from Vitruvian Man and even Joe and Josephine. It is a means for me as an individual to purchase clothes whose measurements are perfect for my body. It fosters, not the appreciation of beauty, nor even the efficient reconstruction of the world, but consumerism and my own subjective will.

How can we keep an eye on the difference between ontic and ontological measurement, and prevent the one from interfering with the other?

One way is to ask about what, if anything, is missing from the measurements delivered by the modern metroscape. We have avatars, and moving avatars, and even moving avatars that reveal how our clothes look from the back and when snagged on twigs that we can consult in the showroom and even at home—so are our clothes fitting any better, or do the entertaining gizmos just make us think they do? Are the tests administered by schools making students smarter and more educated, or just making us think we know how to evaluate education? Is the ability to measure tiny levels of toxins making us safer, or tending to bring about just the opposite, leading us to spend enormous sums of money unnecessarily to eliminate toxins just to make us feel safer? Even in the modern metroscape, measuring does not thrust the rest of human life—the question, "Why do we measure?"—permanently in the background, if we pay attention.

In the modern metroscape, we have to pay more careful attention than ever to the goals we are trying to achieve with measurements, rather than simply to measurements. We have to be more careful to focus on our dissatisfactions, on what measuring does not deliver. And we have to address these dissatisfactions, not by discarding the measures we have and seeking to find newer and better ones, for these, too, will also eventually turn out not to do what we want and eventually need to be renounced, nor by assuming that what we are after lies "beyond" measuring. Rather, the modern metroscape requires us to articulate more carefully what and where our measurements do not deliver.

But the most important way to keep an eye on the difference between ontic and ontological measurement in the modern metroscape is to reflect not simply on how individual acts of measurement are carried out, but on the metroscape itself and what it does to us. Even after we tie the noonday cannon to an absolute standard, we still need to keep reminding ourselves of the human purposes that led us to create it in the first place—and where, if at all, it interferes with any of these purposes. We can do this in part by tweaking ourselves mocktrologically, and by teaching ourselves philosophically, but much more importantly by retelling the story of measurement—reminding ourselves how the modern metroscape came to be, what the alternatives were, why we rejected them, and what we gained but also lost by rejecting them.

ACKNOWLEDGMENTS

This book, like my previous books *The Prism and the Pendulum: The Ten Most Beautiful Experiments in Science* and *The Great Equations: Breakthroughs in Science from Pythagoras to Heisenberg*, grew out of a column I write for *Physics World*, a consistently fun magazine to read and write for. Its editor, Matin Durrani, first got me interested in the subject of this book when he remarked that the two subjects that generate the most feedback and controversy in science magazines are religion and measurement units. Chapters 2, 8, and 12 are based on material that first appeared in the July 2011, December 2009, and March 2011 issues, respectively. I am also grateful to associate editor Dens Milne and to the hundreds of people who responded to my columns about units and measurement. Above all, I am indebted to my editor at Norton, Maria Guarnaschelli, for her many thoughtful readings of early versions of the manuscript, while I struggled to find a way to tell the fascinating and utterly messy story of metrology, and her wisdom in judging when to be patient and when insistent. I am also indebted to editorial assistant Melanie Tortoroli for helping to steer the manuscript through; to Carol Rose, the copy editor; and to Nancy Palmquist. Like all columnists, I rely heavily on colleagues and correspondents for inspiration, ideas, and information, and those who provided significant suggestions, comments, and other kinds of assistance include: Joseph Antista, Peter Becker, Rachelle Bennett, Lindsay Bosch, Edward S.

Casey, Richard Crease, Allegra de Laurentiis, David Dilworth, Joe Dixon, B. Jeffrey Edwards, Patrick Grim, George W. Hart, Robert Harvey, Linda Henderson, Don Ihde, Haiqing Ji, Xiping Jin, Judy Bart Kancigor, Chris Laico, Peter Main, Peter Manchester, John H. Marburger III, Keith Martin, Rita Mazzella, James McManus, Eduardo Mendieta, Hal Metcalf, Kevin Meyer, Lee Miller, Ian Mills, Mark Mitton, David Newell, Seung-Young Noh, Karen Oberlin, Mary Rawlinson, Ian Robinson, Robert C. Scharff, Rhonda Roland Shearer, Jodi Sisley, Michael M. Sokal, Marshall Spector, Ben Stein, Richard Steiner, Richard Stone, Clifford Swartz, Andy Taylor, Barry Taylor, Abebe Tessema, Bob Vallier, Andrew Wallard, Paul Wilby, Nancy Wu, Beth Young, Fan Zhang, and Yajie Zhang. Without the capable help of Alissa Betz, Ann-Marie Monaghan, and Nathan Leoce-Schappin in the department office this manuscript would have been much delayed. Thanks to the Columbia University Library Rare Books Collection. Howard L. Goodman patiently corrected version after version of the pages about China. David Bruner taught me about three-dimensional imaging. David Dilworth taught me about Peirce. Hal Metcalf taught me about spectroscopy. Claire Béchu and Brigitte-Marie Le Brigand showed me the Meter and Kilogram of the Archives. Ruolei Wu helped me navigate around Beijing. Richard Davis let me tap his knowledge as head of the Mass Division at the BIPM; Terry Quinn was equally generous with his extensive institutional memory of the institution. Zengjian Guan, a professor in the Department for the History and Philosophy of Science at Shanghai Jiao Tong University, helped me with the recent history of Chinese metrology. My wife, Stephanie, not only read the manuscript and put up with my difficult and trying work routines but also provided me throughout, as always, with the sounds of surprise. My son, Alexander, once again had to endure my work habits and periodic unavailability, and teach me how to use each new piece of technology. My daughter, India, always made sure I was in the right dimension. My dog, Kendall, was always willing to go for a walk with me when the rest of them got fed up.

NOTES

CHAPTER ONE. VITRUVIAN MAN

1. Daniel Defoe, *The Life and Adventures of Robinson Crusoe* (New York: Greenwich House, 1982), pp. 162–66.
2. Stefan Strelcyn, "Contribution à l'historie des poids et des mesures en Ethiopie," *Rocznik Orientalistyczny* 28, no. 2 (1965), p. 77.
3. Peter Kidson, "A Metrological Investigation," *Journal of the Warburg and Courtauld Institutes* 53 (1990), pp. 86–87.
4. Emily Thompson, *The Soundscape of Modernity: Architectural Acoustics and the Culture of Listening in America, 1900–1933* (Cambridge: MIT Press, 2002).
5. Sabine knew that open windows were an important factor in diminishing the sound, for instead of absorbing the sound like the cushions they transmitted it to the outside and were more "universal" to the extent that the properties of windows were more similar from building to building than were those of cushions. Still, he preferred the cushions because they were far more convenient to experiment with. "It is necessary, therefore, to work with cushions, but to express the results in open-window units," he wrote.
6. See the National Academy of Sciences biographical memoir of Sabine, http://books.nap.edu/html/biomems/wsabine.pdf; also Thompson, *The Soundscape of Modernity*.
7. *The Laws of Manu*, trans. G. Bühler (Oxford: Clarendon Press, 1886), ch. 8, sections 132–34.
8. Eric Cross, *The Tailor and Antsy* (Dublin: Mercier Press, 1942), pp. 31, 86–87.
9. Vitruvius, *Ten Books on Architecture*, trans. H. H. Morgan (Cambridge: Harvard University Press, 1914), book III, ch. I, sections 3, 5, www.gutenberg.org/files/20239/20239-h/29239-h.htm.

10. See Mark W. Jones, "Doric Measure and Architectural Design 1: The Evidence of the Relief from Salamis," *American Journal of Archaeology* 104, no. 1 (January 2000), pp. 73–93.
11. I'm reversing the terminology of the professor of architecture and urban design Robert Tavernor, in *Smoot's Ear: The Measure of Humanity* (New Haven: Yale University Press, 2007), p. 45, who calls creating an artifact standard a "disembodiment." I think it makes more sense to refer to artifacts as embodying measures, and the modern development—replacing artifact standards with natural constants, mediated by technology—as disembodiment.
12. Peter Menzel and Faith D'Aluisio, *Hungry Planet* (Napa: Material World Books, 2005), p. 237.
13. The principal person to promote this term is Hans Vogel, who applied it to the Chinese context in "Aspects of Metrosophy and Metrology during the Han Period," *Extrême-Orient, Extrême-Occident* 16 (1994), pp. 135–52. However, I am vastly expanding the scope of this wonderful term.
14. Herodotus, *The History of Herodotus*, trans. George Rawlinson, book IV, section 196 (Chicago: Encyclopedia Britannica, Inc., 1952), p. 158.
15. Flavius Josephus, *Antiquities of the Jews*, trans. William Whiston, book I, chapter 2, section 2, http://reluctant-messenger.com/josephusA01.htm.

CHAPTER TWO. ANCIENT CHINA: FEET AND FLUTES

1. Interview, Guangming Qiu, July 2, 2010; translator Ruolei Wu.
2. David N. Keightley, "A Measure of Man in Early China: In Search of the Neolithic Inch," *Chinese Science* 12 (1995), pp. 18–40, at p. 18.
3. Mentioned, for instance, by Joseph Needham in *Science and Civilization in China*, vol. 3 (Cambridge: Cambridge University Press, 1959), p. 84.
4. Robert Poor, "The Circle and the Square: Measure and Ritual in Ancient China, *Monumenta Serica* 43 (1995), pp. 159–210, at p. 180.
5. On early Chinese musicology in English, Howard L. Goodman's *Xun Xu and the Politics of Precision in Third-Century AD China* (Boston: Brill, 2010) focuses on the 180 to 300 AD time period. The book's extensive bibliography covers the key sources in Chinese and English.
6. Robert Bagley, "The Prehistory of Chinese Music Theory," *Proceedings of the British Academy* 131 (2005), pp. 41–90.
7. Bell Yung, in B. Yung, E. Rawski, and R. Watson, eds., *Harmony and Counterpoint: Ritual Music in Chinese Context* (Stanford: Stanford University Press, 1996), p. 23.
8. This story is recycled, for instance, in John Ferguson, "Chinese Foot Measure," *Monumenta Serica* 6 (1941), pp. 357–82, at p. 366.
9. Goodman, *Xun Xu and the Politics of Precision*.

10. Goodman, *Xun Xu and the Politics of Precision*, p. 242.
11. Goodman, *Xun Xu and the Politics of Precision*, p. 211.
12. Goodman, *Xun Xu and the Politics of Precision*, p. 176.
13. Goodman, *Xun Xu and the Politics of Precision*, p. 196.
14. Zichu Wang, *Xun Xu dilü yanjiu* [Research into Xun Xu's Di-Flute Tonal System] (Beijing: People's Music Publishing House, 1995).
15. Howard L. Goodman and E. Lien, "A Third Century AD Chinese System of Di-Flute Temperament: Matching Ancient Pitch-Standards and Confronting Modal Practice," *The Galpin Society Journal* 62 (April 2009), pp. 3–24.
16. Jack A. Goldstone, "The Rise of the West—or Not? A Revision to Socioeconomic History," www.hartford-hwp.com/archives/10/114.html (accessed May 4, 2011).
17. Guangming Qiu, *Zhongguo lidai duliangheng kao* [A Study of Weights and Measures Through the Ages in China] (Beijing: Science Publishing House, 1992). Others working on Chinese metrology include Zengjian Guan et al., *Zhongguo jin xian dai ji liang shi gao* [Draft History of Modern and Contemporary Metrology in China] (Shandong: Shandong Pedagogy Publishing House, 2005). See also Zhengzhong Guo, *San zhi shisi shiji Zhongguo de quanheng duliang* [Chinese Weights and Measures: Fourth to Fourteenth Centuries] (Beijing: China Social Science Publishing House, 1993).

CHAPTER THREE. WEST AFRICA: GOLD WEIGHTS

1. On Niangoran-Bouah, see K. Arnaut, "Les Hommes de Terrain: Georges Niangoran-Bouah and the Academia of Autochthony in Postcolonial Côte d'Ivoire," *Kasa Bya Kasa, Revue Ivoirienne d'Anthropologie et de Sociologie*, no. 15 (2009); Karel Arnaut and Jan Blommaert, "Chthonic Science: Georges Niangoran-Bouah and the Anthropology of Belonging in Côte d'Ivoire," *American Ethnologist* 36, no. 3 (2009), pp. 574–90.
2. "The cultural elements of a given society," Niangoran-Bouah decided, "are what they are for the members of that society before becoming the object of study by scientific field workers." He had to change his research program "to discover the reality of those elements through the original context" (Georges Niangoran-Bouah, *The Akan World of Gold Weights*, 3 vols. [Abidjan: Nouvelles Editions Africaines, 1984], vol. 3, p. 12.
3. Niangoran-Bouah, *The Akan World of Gold Weights*, vol. 1, pp. 42–43.
4. Niangoran-Bouah, *The Akan World of Gold Weights*, vol. 1, p. 43.
5. A. Ott, "Akan Gold Weights," in *Transactions of the Historical Society of Ghana* 9 (1968), p. 37, quoted in *Equal Measure for Kings and Commoners: Goldweights of the Akan from the Collections of the Glenbow Museum* (Calgary, Canada: Glenbow Museum, 1982), p. 25.

6. The outcome was Niangoran-Bouah's advanced doctoral thesis, "Gold Measuring Weights of the Peoples of the Akan Civilization," submitted to the University of Paris X (Nanterre) in October 1972, which later formed the basis for his three-volume work, *The Akan World of Gold Weights*; See vol. 1, pp. 22–24.
7. Niangoran-Bouah, *The Akan World of Gold Weights*, vol. 1, pp. 22–25.
8. Niangoran-Bouah, *The Akan World of Gold Weights*, vol. 3, pp. 315, 318.
9. Tom Phillips, *African Goldweights: Miniature Sculptures from Ghana 1400–1900* (London: Hansjorg Mayer, 2010).
10. Interview, Tom Phillips, October 5, 2010.
11. Phillips, *African Goldweights*, p. 97.
12. Phillips, *African Goldweights*, p. 13.
13. Phillips, *African Goldweights*, p. 29.
14. Phillips, *African Goldweights*, p. 30.
15. Phillips, *African Goldweights*, p. 10.
16. Timothy F. Garrard, *Akan Weights and the Gold Trade* (New York: Longman, 1980).
17. Quoted in Garrard, *Akan Weights*, p. 175.
18. Quoted in Garrard, *Akan Weights*, pp. 175–76.
19. Tom Phillips, "Timothy Garrard," *The Guardian*, May 28, 2007.

CHAPTER FOUR. FRANCE: "REALITIES OF LIFE AND LABOR"

1. Interview, Brigitte-Marie Le Brigand, October 11, 2010.
2. Peter Kidson, "A Metrological Investigation," *Journal of the Warburg and Courtauld Institutes*, 53 (1990), p. 71.
3. Whence the abbreviation *lb* for pound, the name *libra* being taken from the balances used with the weights.
4. Henri Moreau, "The Genesis of the Metric System and the Work of the International Bureau of Weights and Measures," *Journal of Chemical Education* 30, no. 1 (1953), p. 3.
5. As in France, so in England: English attempts to establish uniform standards began with King Edgar's declaration in 960 that "the measure of Westchester [the capital] shall be the standard" for his kingdom. But the subsequent history is of one ineffectual attempt after another. William the Conqueror decreed in 1066 that the weights and measures "most trustworthy and duly certified" would be the same as those of his "worthy predecessors," meaning that he would not attempt anything new. The Magna Carta (1215) states that there will be uniformity of measures of wine, ale, and corn, with the first two measures evidently being two different sizes of gallon, the third specified as the London "quarter," the first time an English measure is actually specified in writing. Henry III issued a decree regulating weights and measures

in 1266, Edward I in 1305 (defining an inch as "three grains of barley dry and round"), and Edward III in 1328. A decree of Henry V in 1414 mentions the "troy" pound, a variant of the traditional pound, with 12 rather than 16 ounces; it was named after Troyes, a French town whose market fair was important for British merchants. Troy measures were used principally by jewelers and were lighter than the "avoirdupois" measures used by merchants of heavier goods, to which they were contrasted. Henry VII had an octagonal brass yard standard created, half an inch in diameter and roughly ruled into inches, and several weight standards (1497). All these statutes and actions did little to reduce the diversity of measures in the country.

6. Ronald Edward Zupko, *British Weights & Measures: A History from Antiquity to the Seventeenth Century* (Madison: University of Wisconsin Press, 1977); and *Revolution in Measurement: Western European Weights and Measures Since the Age of Science* (Philadelphia: American Philosophical Society, 1990). Witold Kula, *Measures and Men*, trans. R. Szreter (Princeton: Princeton University Press, 1986).
7. Kula, *Measures and Men*, p. 29.
8. Kula, *Measures and Men*, p. 7.
9. Kula, *Measures and Men*, pp. 19–20. He adds, "If in practice the coexistence of several different measures enabled the stronger party to perpetrate many abuses, then that was a different matter."
10. Kula, *Measures and Men*, p. 22.
11. Kula, *Measures and Men*, p. 21.
12. Kula, *Measures and Men*, pp. 27–28.
13. Quoted in Kula, *Measures and Men*, p. 115.
14. Kula, *Measures and Men*, p. 101.
15. Kula, *Measures and Men*, p. 123.
16. Ilya Ehrenburg, *Memoirs: 1921–1941*, trans. T. Shebunina (New York: World Publishing, 1964), p. 134. Quoted in Kula, *Measures and Men*, p. 12.
17. Steven Shapin, *Never Pure: Historical Studies of Science as if It Was Produced by People with Bodies, Situated in Time, Space, Culture, and Society, and Struggling for Credibility and Authority* (Baltimore: Johns Hopkins University Press, 2010), p. 23.
18. I. B. Cohen, *The Triumph of Numbers: How Counting Shaped Modern Life* (New York: Norton, 2005), p. 44.
19. In Gabrielle Mouton, *Observationes diametrorum solis et lunae apparentium* (Lyons: Matthae Liberal, 1670).
20. One hundred-thousandth of a mille was a digitus, one-millionth a granum, one ten-millionth a punctum. Realizing that this sequence might be difficult to remember, Mouton proposed an alternate sequence: milliare, centuria, decuria,

virga, virgule, decimal, centesima, millesima—though this supposedly easier sequence, with its more latinate designations, only seemed to confuse things.

21. Its new scientific status was revealed by the fact that Picard's measurements would shortly be used by Isaac Newton to develop the law of gravitation.
22. Jean Picard, *The Measure of the Earth*, trans. Richard Waller (London: R. Roberts, 1687).
23. Jean-Antoine-Nicolas de Caritat marquis de Condorcet, *The Life of M. Turgot, Comptroller General of the Finances of France, in the Years 1774, 1775, and 1776* (London: J. Johnson 1787), p. 134.
24. Quoted in John Riggs Miller, *Speeches in the House of Commons upon the equalization of the weights and measures of Great Britain* (London: Debrett, 1790), p. 12.
25. Kula, *Measures and Men*, p. 186.
26. Roger Hahn, *The Anatomy of a Scientific Institution: The Paris Academy of Sciences, 1666–1803* (Berkeley: University of California Press, 1971), p. 163.
27. Quotes from Talleyrand's proposal are from Miller, *Speeches*, pp. 77–78.
28. Guillaume Bigourdan, *Le Système Métrique des Poids et Mesures: Son Établissement et sa Propagation Graduelle* (Paris: Gauthier-Villars, 1901), p. 30. This valuable book reprints many original documents.
29. Cited in Kula, *Measures and Men*, p. 242.
30. Ken Alder, *The Measure of All Things: The Seven-Year Odyssey and Hidden Error That Transformed the World* (New York: Simon & Schuster, 2002).
31. Maurice Crosland, "The Congress on Definitive Metric Standards, 1798–1799: The First International Scientific Conference?" *Isis* 60 (1969), pp. 226–31, at pp. 230–31. Though had Napoleon been in power, Crosland continued, "he would have transformed it into a major occasion for propaganda" and made sure the event was better known among historians of science.
32. Bigourdan, *Le Système Métrique*, pp. 160–66.
33. J. Q. Adams, *Report of the Secretary of State upon Weights and Measures*, (Washington, DC: Gales & Seaton, 1821), http://books.google.com/books?id=G1sFAAAAQAAJ&printsec=frontcover&dq=john+quincy+adams+weights+and+measures&source=bl&ots=eCVpHlzOgq&sig=Yi86GuqX71ZE1QUkkNfmQDzJDGE&hl=en#v=onepage&q&f=false.

CHAPTER FIVE. HALTING STEPS TOWARD UNIVERSALITY

1. Andro Linklater, *Measuring America: How an Untamed Wilderness Shaped the United States and Fulfilled the Promise of Democracy* (New York: Walker, 2002), p. 131.
2. Linklater, *Measuring America*, p. 135.
3. John Playfair. Review of *Base du Système Métrique Décimal*, by Méchain and Delambre, *Edinburgh Review* 18 (January 1807), p. 391.

4. Quoted in U.S. Department of Commerce, *The International Bureau of Weights and Measures 1875–1975* (Washington, DC: National Bureau of Standards), NBS Special Publication 420, p. 8.
5. Miller, *Speeches*, p. 18.
6. Miller, *Speeches*, p. 75.
7. Miller, *Speeches*, pp. 48–49.
8. Zupko, *Revolution in Measurement*, pp. 104–5.
9. Olin J. Eggen, "Airy, George Biddell," *Dictionary of Scientific Biography*, vol. 1 (New York: Scribner's, 1970), pp. 84–87.
10. James Madison, *Letters and Other Writings of James Madison*: vol. 1, 1769–1793 (Philadelphia: Lippincott, 1865), pp. 152–53.
11. Thomas Jefferson, *The Writings of Thomas Jefferson*, vol. 8 (Washington, DC: Thomas Jefferson Memorial Association, 1907), pp. 220–21.
12. Cited in F. Cajori, *The Chequered Career of Ferdinand Rudolph Hassler* (New York: Arno Press, 1980), p. 38.
13. Nathan Reingold, "Introduction," in Cajori, *Chequered Career*, n.p.
14. Cajori, *Chequered Career*, p. 42.
15. Marie B. Hecht, *John Quincy Adams* (New York: Macmillan), p. 263.
16. Hecht, *John Quincy Adams*, p. 264.
17. John Quincy Adams, *Report upon Weights and Measures* (Washington, DC: Gales & Seaton, 1821).
18. Adams, *Report*, p. 120.
19. Adams, *Report*, p. 135.
20. William Appleman Williams, *The Contours of American History* (Chicago: Quadrangle Books, 1966), p. 215.
21 F. R. Hassler, *Weights and Measures: Report from the Secretary of the Treasury in Compliance with a Resolution of the Senate, Showing the Result of an Examination of the Weights and Measures Used in the Several Custom-houses in the United States, &c.*, Document No. 299, 32nd Cong., 1st sess., July 2, 1832.
22. Quoted in Kula, *Measures and Men*, p. 286.
23. Adams, *Report*, p. 55.
24. Jacques Babinet, *Annales de Chimie et de Physique* (1829), pp. 40, 177.

CHAPTER SIX. "ONE OF THE GREATEST TRIUMPHS OF MODERN CIVILIZATION"

1. Robert Brain, *Going to the Fair: Readings in the Culture of Nineteenth-Century Exhibitions* (Cambridge: Whipple Museum of the History of Science, 1993), p. 24.
2. Quoted in Edward Franklin Cox, "The Metric System: A Quarter-Century of Acceptance (1851–1876)," *Osiris* 13 (1958), pp. 358–79, at p. 363.

3. *First Annual Report* (London: Registrar General of England and Wales, 1839), p. 99.
4. Quoted in Cox, "The Metric System," p. 363.
5. Quoted in Cox, "The Metric System," p. 368.
6. John Herschel, *Familiar Lectures on Scientific Subjects* (London: D. Strahan, 1867), p. 432.
7. Herschel, *Familiar Lectures*, p. 445.
8. Act of 28 July 1866 (14 Stat. 339).
9. Many key documents of this story, again, are found in Bigourdan, *Le Système Métrique des Poids et Mesures*.
10. "The Unit of Length," n.a. *Nature*, June 23, 1870, p. 137.
11. "The International Metric Commission," n.a. *Nature*, January 16, 1873, p. 197.
12. "The International Bureau of Weights and Measures," n.a. *Nature*, October 18, 1883, p. 595; the original article was from *La Nature*.
13. Ibid.
14. The text can be found in Louis E. Barbrow and Lewis V. Judson, "Weights and Measures Standards of the United States: A Brief History," Appendix 3, pp. 28–29, U.S. Commerce Department, NIST Special Publication 447.
15. "The Metric System in the United States," n. a. *Nature* May 14, 1896, p. 44.
16. Guangming Qiu, *A Concise History of Ancient Chinese Measures and Weights* (Beijing: Hefei Industrial University Press, 2005), pp. 182, 185.
17. "Gold Coast," *Encyclopaedia Britannica*, vol. 10 (New York: Werner, 1899), p. 756.
18. Frederick Boyle, *Through Fanteeland to Coomassie: A Diary of the Ashantee Expedition* (London: Chapman and Hall, 1874), p. 93.
19. Boyle, *Fanteeland to Coomassie*, p. 389.
20. Henry Brackenbury, *The Ashanti War: A Narrative* (London: Blackwood, 1874), vol. 2, p. 267.
21. Phillips, *Goldweights*, p. 48.
22. Richard Freeman, *Travels and Life in Ashanti and Jaman* (New York: Stokes, 1898), p. 111.

CHAPTER SEVEN. METROPHILIA AND METROPHOBIA

1. "The Metric System," *Scientific American* 42, no. 6 (February 7, 1880), p. 90.
2. The story of U.S. pro- and antimetric activity is told in U.S. Department of Commerce, U.S. Metric Study Interim Report, *A History of the Metric System Controversy in the United States* (Washington, DC: National Bureau of Standards, 1971), Special Publication 345-10.
3. "Shall We Change our Weights and Measures?" *Scientific American* 35, no. 8 (August 19, 1876), p. 113.

4. U.S. Department of Commerce, *A History of the Metric System Controversy*, p. 77.
5. Quoted in S. Schaffer, "Metrology, Metrication, and Values," in *Victorian Science in Context*, ed. B. Lightman (Chicago: University of Chicago Press, 1997), p. 450.
6. John Taylor, *The Great Pyramid, Why Was It Built?* and *The Battle of the Standards: The Ancient, of Four Thousand Years, Against the Modern, of the Last Fifty Years—the Less Perfect of the Two* (London: Longman, Green, 1864).
7. Martin Gardner, *Fads and Fallacies in the Name of Science* (New York: Dover, 1952), has an excellent discussion of "pyramidology."
8. Quoted in H. A. Brück and M. T. Brück, *The Peripatetic Astronomer: The Life of Charles Piazzi Smyth* (Philadelphia: Adam Hilger, 1988), p. 99.
9. C. Piazzi Smyth, *New Measures of the Great Pyramid* (London: Robert Banks, 1884), pp. 105–6.
10. Charles Piazzi Smyth, *Our Inheritance in the Great Pyramid* (London: Daldy, Ishister, 1877), p. 215.
11. C. Latimer, *The French Metric System; or the Battle of the Standards* (Cleveland: Savage, 1879).
12. Latimer, *French Metric System*, p. 23.
13. See Edward F. Cox, "The International Institute: First Organized Opposition to the Metric System," *Ohio Historical Quarterly* 68, no. 1 (January 1959), pp. 58–83.
14. Charles Totten, *An Important Question in Metrology, Based Upon Recent and Original Research* (New York: Wiley, 1884).
15. *International Standard* 1, no. 4 (September 1883), pp. 272–74.
16. Charles S. Peirce, *Philosophical Writings of Peirce*, ed. J. Buchler (New York: Dover, 1955), p. 57.
17. F. A. Halsey and S. S. Dale, *The Metric Fallacy, by Frederick A. Halsey, and The Metric Failure in the Textile Industry, by Samuel S. Dale* (New York: Van Nostrand, 1904).
18. Halsey and Dale, *Metric Fallacy and Metric Failure*, pp. 11–12.
19. Cox, "The International Institute," p. 30.
20. S. S. Dale to T. Roosevelt, January 13, 1905, Samuel Sherman Dale Papers, Rare Book Collection, Box 11, Columbia University.
21. Samuel Sherman Dale Papers, Rare Book Collection, Columbia University.
22. George Kunz to Samuel S. Dale, November 17, 1923, Dale Papers, Box 8, Columbia University.
23. F. A. Halsey to Daniel Adamson, March 17, 1919, Dale Papers, Box 6, Columbia University.

24. W. R. Ingalls to the Members of the American Institute of Weights and Measures, September 24, 1919, Dale Papers, Box 1, Columbia University.
25. Frederick A. Halsey, "Metric System a Failure," *The New York Times*, July 26, 1925, p. 12.
26. S. Dale to F. A. Halsey, August 1, 1925, Dale Papers, Box 6, Columbia University.
27. John Kieran, "Sports of the Times," *The New York Times*, August 5, 1936, p. 26.
28. The U.S. metric controversy story is told in U.S. Department of Commerce, *A History of the Metric System Controversy in the United States* (Washington, DC: National Bureau of Standards, 1971), Special Publication 345-10.
29. Arthur E. Kennelly, *Vestiges of Pre-metric Weights and Measures Persisting in Metric-System Europe, 1926–1927* (New York: Macmillan, 1928), p. vii.
30. Kennelly, *Vestiges*, p. 51.
31. Kennelly encountered some ancient traditions with modern ramifications. In Barcelona, Spain, an important mercantile center, he came across the Colegio Oficial de Pesadores y Medidores Publicos de Barcelona, or the Corporation of Weighers and Measurers of the Commerce of Barcelona. It started as a guild sometime before 1292, the date of the first document in its files, and claims to be the model for similar institutions that started hundreds of years later in other European countries. Its historical artifacts include a large wall crucifix. The custom of the weighers was to make an oath of "fair and true weighing" before they took possession of a ship's cargo to weigh. This illustrates an awareness of the importance of trust and justice in measures that continues today, though in quite a different form.
32. Kennelly, *Vestiges*, p. 51.

CHAPTER EIGHT. SURELY YOU'RE JOKING, MR. DUCHAMP!

1. Paul Valéry, *Aesthetics*, trans. Ralph Manheim (New York: Pantheon, 1964), p. 225.
2. Pierre Cabanne, *Dialogues with Marcel Duchamp* (London: Thames and Hudson, 1971), p. 17.
3. Cabanne, *Dialogues*, p. 59.
4. Marcel Duchamp, *Salt Seller, The Writings of Marcel Duchamp*, ed. M. Sanouillet and E. Peterson (New York: Oxford University Press, 1973), p. 71.
5. Duchamp, *Salt Seller*, p. 160.
6. "April Fool's Joke on Learned Curator," *The New York Times*, April 12, 1925, p. E3.
7. Linda Henderson, *Duchamp in Context: Science and Technology in the Large Glass and Related Works* (Princeton: Princeton University Press, 1998), p. 63.

8. The Web site of the Art Science Research Laboratory is www.asrlab.org.
9. More recently, art historians have acquired a deeper appreciation of the way science influenced this and other art of the early twentieth century, thanks to scholars who have carefully examined the popular literature of the time. These include Gavin Parkinson's *Surrealism, Art, and Modern Science* (2008) and Elizabeth Leane's *Reading Popular Physics* (2008). Henderson's extensively researched book *The Fourth Dimension and Non-Euclidean Geometry in Modern Art* (1983) has been reissued with new material.
10. Kemp Bennet Kolb, "The Beard-Second, a New Unit of Length," in *This Book Warps Space and Time: Selections from The Journal of Irreproducible Results*, ed. Norman Sperling (Kansas City, MO: Andews McMeel, 2008), p. 13.
11. *Physics World*, April 2010, p. 3; B. Todd Huffman, Letter, *Physics World*, May 2010, p. 14; Keith Doyle, Letter, *Physics World*, May 2010, p. 14.

CHAPTER NINE. DREAMS OF A FINAL STANDARD

1. Victor F. Lentzen, "The Contributions of Charles S. Peirce to Metrology," *Proceedings of the American Philosophical Society* 109, no. 1 (February 18, 1965), pp. 29–46.
2. J. Brent, *Charles Sanders Peirce: A Life* (Bloomington: Indiana University Press, 1993).
3. Cited in Max H. Fisch, "Introduction" to Charles S. Peirce, *Writings of Charles S. Peirce, A Chronological Edition*, 8 vols, ed. C. Kloesel (Bloomington: Indiana University Press, 1986), vol. 3, 1872–1878, p. xxii.
4. On this club, see L. Menand, *The Metaphysical Club* (New York: Farrar, Straus & Giroux, 2002).
5. Iwan Rhys Morus, *When Physics Became King* (Chicago: University of Chicago Press, 2005), pp. 253–60.
6. C. Evans, *Precision Engineering: An Evolutionary Perspective*, MSC Thesis, Cranfield Institute of Technology, 1987.
7. François Arago, quoted in *Comptes rendus de l'Académie des sciences* 69 (1869), p. 426.
8. J. C. Maxwell, *A Treatise on Electricity and Magnetism* (New York: Dover, 1954), pp. 2–3.
9. Quoted in Brent, *Peirce*, p. 103.
10. V. Lenzen and R. Multhauf, "Development of Gravity Pendulums in the 19th Century," United States Museum Bulletin 240, Contributions From the Museum of History and Technology, Smithsonian Institution, Paper 44 (Washington, DC: 1965), pp. 301–48.
11. Fisch, "Introduction," p. xxv.

12. Nathan Houser, "Introduction" to Peirce, *Writings of Charles S. Peirce*, vol. 4, 1879–1884, p. xxii.
13. Houser, "Introduction," p. xxviii.
14. Peirce, *Writings*, vol. 4, p. 81.
15. Peirce, *Writings*, vol. 4, p. 269.
16. Peirce, *Writings*, vol. 4, p. 4.
17. "The production of diffraction gratings: I. Development of the ruling art," *Journal of the Optical Society of America* (1949), pp. 413–26.
18. Brück and Brück, *The Peripatetic Astronomer*, p. 175.
19. D. Warner, "Lewis M. Rutherfurd: Pioneer Astronomical Photographer and Spectroscopist," *Technology and Culture* 12 (1971), pp. 190–216. See also the "Biographical Memoir" of Rutherford, B. A. Gould, National Academy of Sciences, books.nap.edu/html/biomems/rutherfurd.pdf.
20. Peirce, *Writings*, vol. 4, p. 241.
21. Daniel Coit Gilman papers, Johns Hopkins University Special Collections, "Peirce" folder, 1883.
22. G. Sweetnam, *The Command of Light: Rowland's School of Physics and the Spectrum* (Philadelphia: American Philosophical Society, 2000).
23. L. Bell, "On the Absolute Wave-lengths of Light," *American Journal of Science* 33 (1887), p. 167.
24. A. Michelson and E. Morley, *American Journal of Science* 34 (1887), pp. 427–30.
25. Brent, *Peirce*, p. 191.
26. In a review of a book by Eduard Noel, an amateur metrological reformer and opponent of the metric system, Peirce ventured into the social aspects of metrology: He was unenthusiastic about Noel's antimetric ideas, but his pragmatic instincts made him doubt its quick acceptance. Considering the way America was parceled into acres and lots, and the way all machinery consisted of parts liable to break and wear out, "and must be replaced by another of the same gauge almost to a thousandth of an inch," Peirce wrote, "[e]very measure in all this apparatus, every diameter of a roll or wheel, every bearing, every screw-thread, is some multiple or aliquot part of an English inch, and this must hold that inch with us, at least until the Socialists, in the course of another century or two, shall, perhaps, have given us a strong-handed government" ("Review of Noel's *The Science of Metrology*," *The Nation*, February 27, 1890).
27. Brent, *Peirce*, pp. 259–60.
28. W. James to J. Cattell, December 13, 1897, in *The Correspondence of William James*, vol. 8, ed. I. Skrupskelis and E. Berkeley (Charlottesville: University Press of Virginia, 2000).

29. Quoted in S. Schaffer, "Metrology, Metrication, and Values," *Victorian Science in Context*, ed. B. Lightman (Chicago: University of Chicago Press, 1997), p. 438.
30. William Thomson (Lord Kelvin), *Popular Lectures and Addresses* (London: Macmillan, 1889), pp. 73–74.
31. A. Michelson, *Light Waves and Their Uses* (Chicago: University of Chicago Press, 1902).
32. Simon Schaffer, "Late Victorian Metrology and its Instrumentation: a Manufactory of Ohms," in *Invisible Connections: Instruments, Institutions, and Science*, ed. R. Bud and Susan E. Cozens (Bellingham, WA: International Society for Optical Engineering, 1992), pp. 23–56.
33. William Harkness, "The Progress of Science as Exemplified in the Art of Weighing and Measuring," *Nature*, August 15, 1889, pp. 376–383, at p. 382.
34. See Schaffer, "Metrology."
35. James Clerk Maxwell, *The Scientific Letters and Papers of James Clerk Maxwell*, ed. P. M. Harman (Cambridge: Cambridge University Press, 2002), pp. 898–99.

CHAPTER TEN. UNIVERSAL SYSTEM: THE SI

1. Charles-É. Guillaume, Nobel Lecture, 1920, http://nobelprize.org/nobel_prizes/physics/laureates/1920/guillaume-lecture.pdf (accessed December 8, 2010).
2. Ludwig Wittgenstein, *Philosophical Investigations* (London: Blackwell, 1958), p. 25.
3. Kula, *Measures and Men*, pp. 267–68.
4. Zengjian Guan, personal communication, December 13, 2010. Zengjian Guan is the coauthor of *Zhongguo jin xian dai ji liang shi gao* [Draft History of Modern and Contemporary Metrology in China] (Shandong: Shandong Pedagogy Publishing House, 2005).
5. Interview, Guangming Qiu, July 2, 2010, translator Ruolei Wu.
6. Beverly Smith, Jr., "The Measurement Pinch," *Saturday Evening Post*, September 10, 1960, pp. 100–104.
7. Edward Teller, "We're Losing by Inches," *Los Angeles Times*, May 15, 1960, p. B6.
8. "New Standard of Meter and Second Set Up," *Chicago Daily Tribune*, October 16, 1960, p. A16.
9. "O, for the Simple Life!" *Chicago Daily Tribune*, October 21, 1960, p. 16.
10. The mole: "the amount of substance of a system which contains as many elementary entities as there are atoms in 0.012 kilogram of carbon 12; its symbol is 'mol.'"

CHAPTER ELEVEN. THE MODERN METROSCAPE

1. Kula, *Measures and Men*, p. 121.
2. Kula, *Measures and Men*, p. 288.
3. F. Gilbreth, Jr., and E. Carey, *Cheaper by the Dozen* (New York: Bantam, 1948), pp. 94–95. Compilations of Therbligs differ slightly, and a modern list contains the following elements: Search, Find, Select, Grasp, Hold, Position, Assemble, Use, Disassemble, Inspect, Transport loaded, Transport unloaded, Pre-position for next operation, Release load, Unavoidable delay, Avoidable delay, Plan, Rest to overcome fatigue.
4. Lawrence Busch and Keiko Tanaka, "Rites of Passage: Constructing Quality in a Commodity Subsector," *Science, Technology, & Human Values* 21, no. 1 (Winter 1996), pp. 3–27, at p. 23.
5. Craig Robinson, *Details*, April 2006.
6. Interview, Joseph Dixon, February 28, 2010.
7. Interviews, David Bruner, Roy Wang, Joseph Antista, June 14, 2010.
8. Kula, *Measures and Men*, p. 12.
9. Charles Dickens, *Hard Times* (Harmondsworth: Penguin, 1969), p. 48.
10. Plato, *Republic*, 603a.
11. Martin Heidegger, "The Question Concerning Technology," in *Basic Writings*, ed. David Krell (New York: HarperCollins, 1993). Given recent controversies over the significance of Heidegger's involvement with the Nazi party, I feel it necessary to address the subject. What Heidegger teaches those who read him, above all, is self-inquiry—to investigate where we stand with respect to others, what we've inherited from the tradition, and what parts of that tradition we want to keep and what to change. That is precisely why those who were weaned on him—who include Levinas, Marcuse, and Habermas—have been his sharpest critics. Indeed, any reader of Heidegger who does not ask the question, Where do I stand with respect to what I am inheriting? has not begun to understand him. The impulse to denounce Heidegger for his Nazi involvement is understandable, but to use that as a justification for questioning his insights is not, and even suspect. How good it makes us far less gifted people feel to denounce the morality of someone as influential as Heidegger! Yet that is not, I think, just a moral cheap shot but unphilosophical. Ethics means posing the question, How can I be better?—which is not a personal question but about one's entire understanding of others—rather than pointing out the faults of others. Those whose idea of ethics is solely to point out the faults of others display not only lack of understanding, but of moral integrity as well. Nobody needs reminding how horrible Nazism was, or

how wrong Heidegger—long after the fundamentals of his philosophy, what's innovative and morally expanding about it, were in place—was in participating in it. Philosophy is so difficult and challenging that the temptation is overriding to find some excuse to sweep books that one does not want to read off the shelves as worthless. Philosophers, and those who appreciate what it means to think philosophically, resist that temptation.

12. Thanks to Robert C. Scharff.
13. This sense is parodied in a scene from the movie *Pulp Fiction*, when Jules asks his friend Brett if he knows why the French call a hamburger with cheese "Le Big Mac." His companion Brett says innocently, "Because of the metric system?" Jules: "Check out the big brain on Brett! You're a smart motherfucker. That's right. The metric system."
14. Herta Müller, *Atemschaukel* (Munich: Hauser, 2009), p. 87.

CHAPTER TWELVE. AU REVOIR, KILOGRAM

1. Terry Quinn, opening remarks, "The New SI: Units of Measurement Based on Fundamental Constants," January 24–25, 2011, Royal Society, London.
2. J. Terrien, "Constants physiques et métrologie," *Atomic Masses and Fundamental Constants 5*, ed. J. H. Sanders and A. H. Wapstra (New York: Plenum, 1976), p. 24.
3. T. J. Quinn, "The Kilogram: The Present State of our Knowledge," *IEEE Transactions on Instrumentation and Measurement* 40 (1991), pp. 81–85.
4. Interview, Peter Becker, January 24–25, 2011, Royal Society, London.
5. In more detail: The mechanical weight of an object ($F = mg$) is matched with the electrical force of a current-carrying coil placed in a strong magnetic field ($F = ilB$), where i is the current in the coil, l is its length, and B is the strength of the field. The device is known as a watt balance because if the coil is moved at speed u it generates a voltage $V = Blu$—and hence, by mathematical rearrangement of the above expressions, the electrical power (Vi) is balanced by mechanical power (mgu). In other words, $m = Vi/gu$. In modern watt balances the current, i, can be determined to a very high precision by passing it through a resistor and using the Josephson effect to measure the resulting drop in voltage. Discovered by Brian Josephson in 1962, it describes the fact that if two superconducting materials are separated by a thin insulating material, pairs of electrons in each layer couple in such a way that microwave radiation of frequency f can create a voltage across the layer of $V = hf/2e$. The resistance of the resistor, meanwhile, is measured using the quantum Hall effect, which describes the fact that the flow of electrons in two-dimensional systems at ultralow temperatures is quantized, with the

conductivity increasing in multiples of e^2/h. The voltage, V, is also measured using the Josephson effect, while the speed of the coil, u, and the value of g can also be easily obtained.

6. About 1978, Kibble and his colleague Ian Robinson built a balance at the NPL. By the end of the decade they had made measurements to a few parts in 10^7 and these results were combined by international committees with other results from around the world and then used to fix the conventional value of the Josephson constant $K_{J\text{-}90}$, which led to agreement among voltage measurements worldwide. Meanwhile, scientists at the NBS in the United States were also operating a balance to improve knowledge of the SI volt. The U.S. scientists named it a watt balance, for power is measured in watts, and the name stuck.
7. T. J. Quinn et al., "Redefinition of the Kilogram: A Decision Whose Time Has Come," *Metrologia* 42 (2009), pp. 71–80.
8. B. Andreas et al., "Determination of the Avogadro Constant by Counting the Atoms in a ^{28}Si Crystal," *Physical Review Letters* 106, 030801 (2011).
9. M. Planck, "Über irreversible stralungsvorgänge," *Annalen der Physik* (1900).
10. "The New SI: Units of Measurement Based on Fundamental Constants," January 24–25, 2011, Royal Society, London.

EPILOGUE

1. Steven Crowell, "Measure-taking: Meaning and Normativity in Heidegger's Philosophy," *Continental Philosophy Review* 41 (2008), pp. 261–76.
2. Friedrich Hölderlin, "In lovely blueness . . . ," in *Hölderlin*, trans. Michael Hamburger (New York: Pantheon, 1952), pp. 261–65, at p. 263.
3. Flavius Josephus, *Antiquities of the Jews*, book 1, chapter 1, "The Constitution of the World and the Disposition of the Elements," www.earlyjewishwritings.com/text/josephus/ant1.html.
4. Gary Wolf, "The Data-Driven Life," *The New York Times Magazine*, April 28, 2010.
5. According to a Web site about the film, "Rosler's distanced depiction of the systematic, institutionalized 'science' of measurement and classification is meant to recall the oppressive tactics of the armed forces or concentration camps, and to underscore the internalization of standards that determine the meaning of women's being" (Rosler Web site: www. eai.org/Title.htm?id=2599).

ILLUSTRATION CREDITS

page 27: Courtesy of the Ashmolean Museum, University of Oxford.
page 27: T. E. Rihll.
page 28: Library of Congress.
page 38: Courtesy of Zhou Xian, dean of the Institute for Advanced Studies in the Humanities and Social Sciences, Nanjing University, and Howard L. Goodman.
page 41: Robert W. Bagley.
page 44: Author's photo.
page 50: Zhao Wu.
pages 58 and 64: Photos by Heini Schneebeli, courtesy of Tom Phillips.
page 71: French National Archives.
page 74: Musée des Arts et Métiers.
page 86: Paris Observatory.
page 89: Library of Congress.
page 95: Rachelle Bennett.
page 110: Library of Congress.
page 116: Library of Congress.
pages 136 and 138: © BIPM.
page 140: NIST.
page 157: Columbia Library Rare Books Collection.
page 170: © 2011 Artists Rights Society (ARS), New York / ADAGP, Paris / Succession Marcel Duchamp.
page 178: Jim McManus.
page 185: National Oceanic Atmospheric Administration.
pages 193 and 197: Pictures by Stefan Kaben, courtesy of the Smithsonian Institution.

page 229: *The Measure of Man & Woman: Human Factors in Design*, revised edition, by Alvin R. Tilley and Henry Dreyfuss Associates. Copyright © 2002 by John Wiley & Sons, New York. Reproduced with permission of John Wiley & Sons, Inc.
page 237: Courtesy of David Bruner and $[TC]^2$.
page 241: Courtesy of Rita Mazzella.
page 254: © BIPM.
page 257: Physikalisch-Technische Bundesanstalt.
page 259: Richard Steiner, NIST.
pages 261, 263, and 264: Ian Mills.

INDEX

Page numbers in *italics* refer to illustrations.

ABOUT THE AUTHOR

ROBERT P. CREASE is a philosophy professor and chairman of the Department of Philosophy at Stony Brook University. He writes a monthly column, "Critical Point," for *Physics World* magazine. He is a Fellow of the Institute of Physics (IOP) in London and of the American Physical Society (APS) in the United States. He edits the APS's Forum on History of Physics newsletter. He has written, cowritten, translated, or edited over a dozen books. His previous books include *The Great Equations: Breakthroughs in Science from Pythagoras to Heisenberg*, *The Prism and the Pendulum: The Ten Most Beautiful Experiments in Science*, *Making Physics: A Biography of Brookhaven National Laboratory*, *The Play of Nature: Experimentation as Performance*, and *The Second Creation: Makers of the Revolution in Twentieth-Century Physics* (with Charles C. Mann). His translations include *American Philosophy of Technology: The Empirical Turn* and *What Things Do: Philosophical Reflections on Technology, Agency, and Design*. He lectures widely, and his articles and reviews have appeared in the *The Atlantic*, *The New York Times*, *The Wall Street Journal*, *Science*, *New Scientist*, *American Scientist*, and other scholarly and popular publications. His Web page can be found at www.robertpcrease.com. He lives in New York City with his family.